I0489767

Inertial Propulsion; the dynamics of the third law in energy form

Inertial Propulsion; the proof of efficiency, kinematic proofs, mechanical energy proofs, free design Software and much more©

Edition:2016-5, 8.1

ISBN-13:
978-1533049759

ISBN-10:
1533049750

Author: Gottfried J. Gutsche, Web site: www.realautomation.ca

Email: info@realautomation.ca

Please use the above Email address to order the free IP Design Software

The complete older publication in lecture format and the example inertia drive info is also available from www.mindbites/series/1278/

Patent protected: US 8491310
Patents Pending:
US 11/544,722 , US 12/082,981, US 12/932857, US 12/802,388, CA 2,526,735.
US 14/120031

Table of Content

Foreword

The objective of this publication is to present a comprehensive practical study to determine the vitality of Inertial Propulsion employing time tested realistic successful engineering analytical techniques. From multiple practical examples the study proceeds to the identification of the fulfillment path toward its full potential. The analysis completes with a practical, obvious, formal and numeric mathematical proof using Newton's own math derivation method for the pendulum and using his pictorial vector analysis. This Proof is verified with rigorous pendulum tests measurements and is extending the known proportional relationship of the angular spin frequency $(2\pi/t)$ in relation to the straight line impulse reflections from a full cycle to 1/4 angular turns $(\pi/2)$ of the arc motion cycle. This proof has been presented as a contribution to the 2013 WSEAS conference on Mechanical and Mathematical modeling. Within the present study, frequency modulated mechanical oscillators working mutually-reciprocally and operating in the complex Cartesian grid are examined for a capacity to develop a cyclic repeating net self-contained propulsion thrust impulse force drive in a predetermined direction in correlation to 1/4 turns. The self-contained net thrust impulse drive means that the net product of average force multiplied by the cycle time is larger in the direction of a vehicular motion. The impulse is internally generated without traction or propellant expulsions. The Reality check with quantifiable Pendulum Tests measurement Videos are available from **www.mindbites.com/series/1278/**

The main objective of this publication is to describe the author's Inertial Propulsion Devices in a practical, realistic and in a verifiable form, packaged in a coffee table book format; presenting the formulas, methods and proofs used to engineer the device. The technical improvements necessary to reach the absolute optimum thrust performance of the presented technologies are also discussed in detail. The practical established existing mechanical construct examples, used within the publication as stepping stones, have an indisputable level of certainty in comparison to purely abstract physics thinking. At the same time, the publication represents a thorough scientific investigation comprehensible by a large general audience, school and media personnel with firm knowledge of college math and physics and it appeals to readers with keen interest to investigate new technologies.

The genesis of this publication was urged along by sounding board type conversations and book reviews by many PHD level expertise individuals to name a few: Prof. Dr. C. Provatidis, Prof. D.E.Simanek, Dr. D. Allen, R. Blackburn, J. Mackie, Dr. R. Anderton, P. Priedemuth, Dr. Glen A Robertson, Dr. DeMees, Dr. D. Tombe, Professor P. Prochnow.

The presented calculations for the engineering of the propulsion device use the units of Kgf and N for force, kinetic energy in units of Kgfm and Joules. To illustrate these forces and the associated kinetic energies, the one Kgf is simply the **force** a 1Kg package presses to the ground in Paris France. The **force** of a 1 Kg package to the ground is only fractionally different in the reader's location.

The **N** (Newton force) is a measurement of mass reluctance (inertial resistance) to change in motion. One Newton **force** accelerates 1Kg mass to $1m/s^2$ (one meter per second$^{2)}$). The Earth gravity accelerates 1Kg mass to $9.8m/s^2$. The 1Kg mass is then defined as $1Nforce*1s^2/1m$. If all this sounds too complicated let's just say that 1 Kgf is about 10 **N**. The length, denoted as "**s**" in equations, is conveniently reproduced with a meter measuring tape and the product of Kgf multiplied by the meter is the kinetic energy of 1 Kgfm = 9.81 Joule (the force of 1 Kgf exerted over 1 meter distance = 9.81 Joule). The One (1) Kgfm energy is about the electrical energy of 0.003-Watt hour on the watt meter.

The measure for the frequency of rotation is **RPM** (revolution per minute) and the angular velocity ω is used to illustrate the cycle frequency in radians per second (2π/second). RPM is used for the car dashboard tachometer, while angular velocity is used for circular inertial mass motions. It might be considered old fashion to use Kgf and RPM, but a technical person will appreciate N_{Newton} and angular velocity ω. For everyday reality of these measurements a complete layman will appreciate Kgf and RPM.

This publication uses references selected on the highest merit, certainty, quality and reality based on practical time proven examples. The Engineering reference: Kurt Gieck Engineering Formulas 7Th Edition-section L1-L10.
For verifying mechanical examples this publication uses: Schaum's 3000 Solved Physics Problems.
For Calculus problems Schaum's 3000 Calculus problems: 20.60, 31.26, 31.31.
For complex control systems: Schaum's Feedback and Control Systems by DiStefano.
Physics for science by M. Browne.
Mechanics presented in a new form by Heinrich Hertz.

For simplicity, premier certainty and clarity the use of differential calculus expressions of parameter instantaneous delta/delta rate of change (snapshot changes, derivatives/slopes) are only used when the physics function is fully proven. This is because of the uncertainty and complexity of how the instantaneous localized rate of change (slope) varies within the propulsion working cycle time-frames by the applicable physics/math functions. In contrast, the primary rule of the slope of the secant line, the mean value theorem analysis method is compared to the "sum of snapshot of changes" method; also referred to as the calculus of changes.

The mean value is describing the average slope and the integral of the parameters magnitude of the Y-axis-gain and X-axis-gain changes spanning the propulsion cycle. This principle is also commonly referred to as: "Rise over run". The word "gain" is used to indicate the change (change=Gain=Rise) and is for the entire cycle; infinitesimal small delta is denoted as dV, for example:

"Velocity$_{gain}$/time" is acceleration; the "Velocity$_{-gain}$/time" is negative gain and is de-acceleration. The secant line rule describes the average rate of changes over the entire propulsion cycle. For example:

"Speed$_{average}$, m/sec.= Displacement$_{gain}$/Time$_{duration}$" is the mean value of speed.

The following formula structure is used: First the parameter name is called, for example: "Speed, or Force". Then after the first comma the mode of the parameter, for example: "Force, average or speed, peak or speed, steady" is called. Then after the third comma the measurement scale is called, for example: "Force$_{average}$, N, or Force, average, kgf" is called.

This publication always uses the meter for displacement, the second for time and Kg for inertial mass, these parameter measurements are generic, for example: "Force, average, N = mass * Velocity, gain / time, duration". Here meter, seconds and Kg mass are generic.

The * is used as the multiplication operator; instead of the square root sign the fractional exponent $x^{1/2}$ is used, like $\sqrt{X}=X^{1/2}$, the square root is factual 1/2

If the reader is unfamiliar with the following math concepts it is recommended to review the following References:
www.en.wikipedia.org/wiki/Mean_value_theorem
www.wikipedia.org/wiki/integral
www.ehow.com/how_4963946_calculate-average-force.html
www.en.wikipedia.org/wiki/vector_space
www.en.wikipedia.org/wiki/Feedback#In_mechanical_engineering/
For Rotational Dynamics and the Flow of Angular Momentum:
www.physikdidaktik,uni-karlsruhe.de/.../rotational-dynamics.pdf
For proof of uniform versus non-uniform angular motion:
www.farside.ph.utexas/teaching/301/lectures/node89.html
The priority and mechanisms described by this publication are protected by Patent and patent applications: US 8491310, US 11/544,722 , US 12/082,981, US 12/802,388, CA 2,526,735, US 14/120031.

Assumptions

The processes and the methods of the Inertial Propulsion systems presented herein are based on a combination of known laws of physics, therefore, have the same inherent assumptions and limitations as these known laws of physics. However, the assumptions of the mass motion laws are examined to determine how these assumptions are consistent with the experimental reality of the measured thrust magnitude of the Inertial Propulsion Drive presented. In summary: The following physics laws and their inherent assumptions apply. The thrust magnitude derivation presented herein has been verified with exhaustive pendulum and ramp climb measurements of working models, while the presented accelerated reference frame motion derivation reality has only undergone rudimentary experimental verification at this time. The postulates presented here are based on the following assumptions inherent in the various Physical laws:

1) The laws of physics hold together with their observations within an isolated moving inertial platform frame and also in the IP machines presented and these laws are assumed to be totally independent of the platform steady velocity magnitude; since the platform is an inertial system.

2) The law of the kinematic quadratic escalation of kinetic energy content for the uniformly accelerated inertial mass motion is assumed to apply; is also vector-kinematical proven and experimentally verified to be true.

3) The law of the relationship of the angular inertial point mass motion kinetic energy quantity (for a pendulum type device) direct proportionality with the product of the centrifugal force multiplied with one half the radius distance (the angular work / energy theorem) is assumed to apply; is also vector-kinematical proven and is experimentally verified to be true.

4) The law of conservation of kinetic energy and energy in general is assumed and experimentally shown to be the premier conservation law of the presented devices; within the presented isolated mechanical systems, no energy is gained nor lost, wherein the total energy sum includes the kinetic energy content of the centre of inertial mass CM of the whole aggregate inertial mass system of particles, and this total sum is verified to be constant.

5) The laws governing the vector-kinematic conversion of cyclic variable rotational inertial mass motion into a straight line mass motion in the complex Cartesian rotational vector grid causing cyclic kinetic energy avalanche discharges are examined; it is also vector-kinematical proven and experimentally verified to be true, wherein Newton uses the kinematic vector type term "combined motion reflections" for the conversion of rotation to straight linen motion applying to his exception to his third law, this indicates his probable prior knowledge of this relationship.

6) The law of conservation of momentum for straight line mass motions, for angular mass motions and for the conversion of rotational to straight line mass motions is examined in view of its very first original root cause and its final disposition into a motionless state, is vector-kinematical proven to be always the conservation of transient momentum in its (kinetic) energy form; it is assumed, there is no inertial mass motion change possible, within machines, without the very first original root cause of energy in form of chemical, solar, wind or nuclear energy.

7) The law of acceleration of an aggregate inertial center of mass (CM) with cyclic unbalanced impulses in the energy form is assumed to apply to the mechanism presented; is also vector-kinematical proven and experimentally verified to be true.

8) The law governing the directional reversibility of Physics (see Lathwait) Principles applies within the mechanics presented; is also vector-kinematical proven and experimentally verified to be true.

9) The law of equal reciprocal reaction to the action of an impulse (third law) is examined in view of the preceding 8 energy based laws; for reasons of symmetry with the physical vector-kinematic magnitude calculation derivation the third law work / energy counterpart derivation is presented, is assumed to apply for the conversion of rotational motion into straight line inertial mass motion; it is also dynamical proven and experimentally verified be true.

The historical development of physics in view of Inertial Propulsion

Physics is the study of matter, energy, space-displacement within the passage of time and the root causes of how they interact in nature. The study results are quantifiable-physical experimental proofs of theses interactions. Throughout this publication the study of physics is used to investigate mechanisms for their ability to provide Vehicular Inertial Propulsion resulting in a quantifiable-physical experimental proof supporting these capabilities.

The very beginning of humanity was marked by the use of arm launched projectiles. The arm muscle thrust to leverage length instilled a fast motion into these objects giving them the ability to do destructive work.

The birth of civilization was in part helped by establishing schools for categorizing, questioning and rationalizing physics phenomena, not for survival, but more for satisfying curiosity, mental exercise and encouraging reasoning. By this time, the ability of projectiles to perform destructive work was categorized as angry motion. This was thought of as a **very** special useful state of motion in comparison to docile pedestrian motion. This knowledge of violent motion was already used by the Greeks and Romans to design large machines of war employing tensioned mechanical lever contrivances capable of delivering very high thrusts for accelerating projectiles to a terrifying violent motion performance, these machines were called the "Ballista". The reason for the diminishing ability to deliver the violent motion of the Ballista projectiles over very long distances was misconstrued as a natural decaying property of the projectile material; lead had the best violent motion performance over distance than wood, this was thought of as an aging effect.

Greek soldiers already used a form of half Inertial Propulsion by orbital rotation of hand held weights to improve their ability to jump over obstacles, wherein the mass acceleration had its backrest against the ground.

More than 2400 years ago, marked the very beginning of quantifiable mathematical and physics thinking when an archetype scientist named Archimedes used the first recorded quantifiable experimental proof of flotation-displacement to convict a crooked goldsmith. This is the moment in history where we accepted the logical truth of a physical experiment as absolute. Archimedes performed many ground breaking breakthroughs in mathematics, mechanics and defined the volume of spheres in relation to cylinders at exact $2/3:1$ ratio by mathematically sidestepping the pesky π constant. Archimedes also made a pronouncement applying to the scalability of physics principles, he said: "Give me a fixed point to stand on and I will move the Earth". This statement seems to tell us that there must always be a fixed point to move an object of substance; therefore, the notion of inertial propulsion ought to be rejected by thinking in terms of levers, pulleys, ropes and minimal inertial mass velocities. This requirement of a fixed point will be traversed by the presented IP using the principle of phased mutual and reciprocal reflection action by the force exertions of rotating masses; this is also referred to as a flailing flywheel motion or the mechanically simulated throw into empty space.

The beginning of the Renaissance

During the Renaissance there was a return to exact quantifiable analytical scientific thinking invented by Archimedes. This was brought about by the discontent with the large discrepancy of authoritarian teachings in comparison to the observable reality of nature. There were three obvious pressing discrepancies observed in nature: The observed yearly seasons to the prescribed calendar seasons, the observed planetary motion in comparison to the prescribed planet motions and the observed free fall motion of objects to the prescribed free fall motion of objects was a source of discontent. The discontent was prompting scientists to logically investigate and prove physics principles with incontestable exact quantifiable experiments. The exactness, obviousness and simplicity of the experiments were vitally important to avoid the wrath of the authorities by gaining fast popular citizen support for the teachings of these experiments. Presenting unsupported speculative theories had in those days a detrimental impact on one's life. In particular, the subject of inertial mass motion was brought into the forefront of science by a surprisingly simple doctrine shattering experiment by Galileo.

Galileo rolled cannon balls down an inclined board having **equal** (repetitious) **spaced** notches inscribed. The clicking noises made by the cannon ball hitting the **equal spaced** notches were having an ever shorter time interval and an ever higher pitched sound as the ball descended; indicating a non-uniform temporal (time) behavior of this inclined notched board system. The click time follows in simplified form a 1/1, 1/3, 1/5, 1/7, 1/9th fractional time interval progression, the fifth time interval is 1/9th fraction of the first time interval. The dropped height was causing an increase in cannon ball speed following a diminishing total returns progression of a 1,3,5,7,9 multiplier in respect to an equal distance interval, wherein the speed from the first notch is multiplied by 9 to obtain the speed at the fifth notch. Galileo presented a lengthy math solution /proof to the notched board experiment in the form of a complicated word problem requiring very high disciplined thinking skills. This complicated verbal math construct was in part responsible for the trouble he got into with the authorities; very few authoritarian figures did appreciate his deep thinking skills, instead his writing was misconstrued as devil talk. Following is a picture of a modern version of Galileo's notched board with bell-ringers at adjustable positions instead of notches. The bell-ringers can be positioned for equal spacing giving an increasing beat frequency and progressive larger spacing giving a repetitive steady beat frequency.

With today's knowledge we describe the action between consecutive notches by saying: The depleted potential energy in form of dropped height is not lost, it is transferred into inertial mass motion kinetic energy; this is today referred to as the Lagrangian-Hamiltonian principle. If this concept is unclear, let's say that the dropped height of one notch distance is proportional to the monetary **Work / Energy** cost (the labour cost) to lift the cannon ball inertial mass back up one notch at the dropped speed. Additional, if we let the ball run out its course on a flat horizontal (level) surface covered by a velvet cloth, then double the number of dropped notches will double the run out distance; or triple the number of doped notches will be triple the run out distance. Galileo also said: The total dropped height is proportional to the square of the acquired canon ball speed magnitude. Galileo's word math and his total volume of research, including the motion of the Earth, were sufficient to refute Archimedes claim of displacing the assumed stationary Earth within a lifetime of labor employing pulleys and ropes. From there, a quest developed to improve Math-Algebra tools to better describe the strange diminishing returns behavior of Galileo's experiments. Galileo also proved with the horizontal (level) run out distance of the cannon ball on a velvet cloth surface that friction was causing the decline in projectile motion over a distance not "aging" of the "violent motion" and the run out distance was decreasing proportional with the friction magnitude. Additional, he proved that the run out distance was in proportion with the number of notches dropped on the declined board; for example, 10 or 5 or 1 number of notches dropped was proportional to 10 or 5 or 1 equal spaced runout distances. With this experiment he gave a pre-notion of the kinetic energy and conservation of energy theory of today. This is the first recorded analysis of the first original root cause of inertial mass motion in combination with the first demonstrated need to quantize inertial mass motion in relation to friction / energy losses. The development of science was now an unhindered popular gentlemen's sport of Diplomats, Mayors, Merchants, Tailors and Courtiers. Science topics were being discussed in coffeehouses, parlors and royal courts, discussing exciting new experiments proving previously thought impossible concepts ranging from the vacuum pump creating a space of nothing, the nature of electricity, the gunpowder motor, to the possibility of building chariots lighter than air. These experiments, in particular the vacuum experiment, were viewed as so important that very top Royal State echelons came to view the first elaborate staged vacuum experiment by the Mayor of Magdeburg Anno 1654. This marks the very first demonstrated top European Royal Support for intellectual freedoms; this is twelve years past Galileo's tenure in jail for speaking proven truth, his subsequent punishment by confinement to house arrest and death; all this because of his large volume of scientific work based on absolute proof. In antiquity we put a crooked gold smith to death because of a newly invented ability to prove his guild beyond a reasonable doubt. Is the dogmatic era still lingering into the modern day where the "beyond a reasonable doubt because of proof" is not the most urgent lawful criteria of our leading authorities?

The extrapolation of technological proofs into the science of the future was never an urgency of the political appointed gatekeepers of scientific truth; this was painfully evident in the technological wakeup calls from the "deemed impossible" human powered flight of the Wright brothers and the foreign "Sputnik" space vehicle launch. The Wright technology was first mathematical proven positive by O. Chanute and the Sputnik technology was first derived by K. Tsialkovsky, H. Oberth, experimentally proven past the speed of sound by H. Goddard and patented within his 214 US patents which were mostly ignored or officially dismissed as unworkable dead ends. The present patented, formally physical five way proven and experimentally working IP principle will probably also endure the same passage, from the deemed unworkable dead end file, into the foreign wakeup call category. Perhaps, in view of this endlessly repeating technological wakeup call scenarios, is it time to broaden the truth base of science to include voices from inventors reaching proof positive results like the Wright brothers, Goddard etc. or more than 90 IP inventors, presenting formally proven successful engineering constructs? This IP denial constraint imposed by authority appointed gate keepers of scientific expediency appears incongruent with the equality concept of democracy and is contrary to our intellectual freedom laws if inventors are constrained, in any way, expressing their formal proofs to government officials coerced to unconditional IP rejection. The next picture depicts the moment in history where central European popular culture gained intellectual freedom; it is the famous Vacuum Experiment of 1654. This principle, or its reverse pressure principle, it is the gas engine energy principle and subsequently allowed us to examine the inner workings of atoms. This freedom of expression spread from here throughout the world. The group of gentlemen in the background are the Emperor of the Holy Roman Empire, the King of Saxony and the mayor of Magdeburg:

The Age of Enlightenment

The fertile new era of popular intellectual freedom produced two prominent continental European Scientists, G. Leibniz and C. Huygens, first being introduced in the fashionable society of Paris. Leibniz and Huygens, working with reciprocal cooperation, correctly identified Galileo's notched board experiment to be congruently (proportionally) related to the escalating performance of projectile motions hurled by Ballista type machines of war delivering the escalating ability to do destructive work against castle walls. This ability was demonstrated with the inclined equal distances notched board; when 5 notches dropped height were deforming 5 stacked tin wires at ounce, then 10 notches dropped height would deform 10 stacked tin wires exactly the same amount, a proportional relation. They renamed the "special angry state" of the mass motion velocity quantity to do destructive work "Vis Viva": Wherein the Vis Viva is thought off as the living Force contained within any inertial mass velocity quantity. This was a logical selection also found in the Greek language; in the French language the Vis Viva is "Energique"= the energy within inertial mass motion, this is congruent with all European languages: In English=energetic, in German=energish, in Russian=energia, etc. Accordingly, these European Barbarians had already adopted a clear single word language concept of inertial mass motion and the destructive ability of high speed inertial mass motion. At last, Huygens and Leibniz assigned an easy verbal, mathematically quantifiable and continual scalable three variable one unknown formula reality to this inertial mass motion eternal mystery by postulating that the energetic potency (destructive) content of inertial mass motion is the

Vis Viva,

a proportional relation to the projectile velocity quantity squared V^2.

Wherein the proof is the gravitational Vis Viva:

$$E_{nergetic} = m_{,mass}\, g_{,gravitational,acceleration}\, h_{,height} = mV^2/2$$

Therefore, the Vis Viva is:

$$\text{Vis Viva} = mass*2distance^2/time^2 = mass*acceleration*distance = \mathbf{mgh}$$

This is reveals the **fundamental** kinematics $V^2 = 2*acceleration*distance$ correlation also contained in the differential equation for oscillations: $f(\mathbf{a}) = d^2s/dt^2$ and it reveals the heavy lifting of the inventing individuals. This inertial mass acceleration*distance, Force*distance, work performed to Vis Viva relation was easily reproduced, quickly verified to a high degree of accuracy and quickly thankfully accepted by all prominent scientists across continental Europe as the base fundamental principle, as expressed within many exchange of letters. Huygens, a member of the Royal Society in London, reserved the priority of this discovery anno 1667 in a letter and a visitation presentation to the Society. Huygens Vis Viva principle also easily fell into obvious logical congruence with Robert Hooke's prior elasticity law of 1660, wherein the Vis Viva energy calculations finally provides the spring force inertial mass motions speed gain. Huygens successfully used the Vis Visa $E_{nergique}$=**mgh** principle, **in form of average force over half the motion distance**, to obtain the quantities of inertial mass motion within mechanical motion constructs. He used the average force over halve the motion distance to solve the pesky isochronous cycle time of the pendulum and the gravitational inertial mass impulse collision from a position of height; these "Huygens Formulas" have not changed, they are still used today in their original form. The Vis Viva principles finally allowed the solving of many more Machine design problems, for example: It solved the calculation of the water wheel transmission driven flour mill, the water wheel transmission driven hammer mill and the ballistic flight of the crossbow arrow and the cannon ball by applying the mean values applying to the $E_{energetic}$=**mgh** magnitude. Importantly, it finally allowed the quantity / magnitude **verification** of the popular folk wisdom notion that **speed kills**. Huygens documented his work in a stream of far reaching important papers: the "Motu Percussione: A.D. 1656", the "Centrifuga: A.D. 1659", and the "Oscillatorium: A.D. 1673" laying the foundations of quantifiable rotational dynamics based on potential energies **mgh** transferring into motion quantities $mV^2/2$. With these scientific papers Huygens first used the Vis Viva principle in conjunction with the geometric pictorial vector type principle of motion quantities having both magnitude and direction to invent and proof the energy symmetry between the straight line Force of kinetic energy in relation to the centrifugal force; he proofed the Vis Viva relation to the moment of inertia and the principle of motion in angular form. By cooperating and helping Rene Descartes presented pre-notions to future developments of Lagrangian, L. Euler, R. Hamiltonian and H. Hertz mechanics. Huygens and Leibniz maintained a lively scientific correspondence and visits discussing these principles openly in great detail, reciprocally correcting-helping each other in an amazing collegial manner without any fear of losing intellectual property.

When the subject of calculus Math was discussed between Huygens and Leibniz, Huygens was less enthusiastic to the idea, pointing to his success in solving many difficult mass motion problems for clocks and machines with his Vis Viva principle in the mean value **displacement domain**; wherein displacement domain means: inertial mass motion is vector-cinematically derived with the displacement "**S**" in the product of force multiplied with the displacement (mgh, h=s) for the Vis Viva; for determining the force, the displacement is placed in the denominator in a fraction of the squared velocity $F=mV^2/2h$. Here we must point to the difference between calculus based inertial mass motion and mean value based calculation. Calculus Math is concerned with changes in magnitude, the gain of a value in small increments; while the mean value is concerned with averages of the magnitudes of cyclic motion; for example the average speed=distance/time; the Vis Viva is the mean value of the energetic force.

Leibniz wrote the first book for teaching calculus Math in A.D.1684 by inventing the modern **dx/dt** notation for infinitesimal small motion changes per time interval (the time derivative) symbol-system and the procedures for the theorem of calculus to arrive at the average values exerted by these quadratic functional systems; he summarised a set of algebraic exponent rules making cumbersome word logic unnecessary. He showed with calculus expressions that within the Vis Viva $E_{energetic}=mgh$ formula, a steady variable force (a force having a straight line incline or decline in respect to the distance traversed) applied over the distance **h** having a mean value force at the half distance **h/2,** then the mean value force still returns the true energy quantity when multiplied by the total accelerated distance; this is making the Vis Viva principle universally applicable to machine design analysis wherein forces are steadily variable (straight line sloped) in respect (to the denominator in the displacement) to equal incremental motion distances. Leibniz also solved in **1686** the mystery of **in-**elastic collisions by proving the extension of the principle of the conservation of the Vis Viva magnitude into inertial mass collisions; Leibniz based this principle on the frictional heat loss between each tiny particles of the malleable material. Leibniz followed Huygens into the inventing field by designing in A.D. 1673 the first hand operated full **four function** mechanical calculator, and he invented an efficient method for calculating **π,** allowing the fast calculation of **π,** the calculation of trigonometric tables and many more math constants quickly to a very high degree of accuracy.

The most prominent, successful, liked and celebrated theoretic physics scientist of the new era of Enlightenment was Newton. It was Newton who had the great, profound and far reaching idea to remove the strange diminishing velocity gain behavior of the **mgh** mass motion by reformulating in **1687** Galileo's notched board math word problem into uniform (repetitious) time intervals, **INSTEAD** of the equal **distance** intervals of the **20** years prior in **1667** invented Vis Viva principle: $mgh_{VisViva}/h=mg=ma=F$. When analyzing a straight line displacement inertial mass motion velocity in equal repetitive (isochronous) time interval progression **without direct** displacement length **(h)** considerations the diminishing returns temporal behavior we have seen in the displacement analysis disappears and a beautiful simple dynamic proportional relationship between force, acceleration, mass and time is presented. This proportional relationship is very easily understandable as a simple verbal problem and as a simple two variable one unknown formula construct:

$$\mathbf{a}_{acceleration} = \mathbf{F}_{force} / \mathbf{m}_{mass} ;$$

Expressed in words: The **higher** the force or the **lighter** the mass magnitude the higher the acceleration!

Then, the product of **ma** is Force:

$$\mathbf{F}_{Force, average} = \mathbf{ma};$$

Accordingly, in kinematic-vector terms acceleration is:

$$\mathbf{a}_{acceleration} = \mathbf{V}_{gain} / \mathbf{t}_{time, duration, of\ gain} ;$$

important: the time duration-delay for the $\mathbf{V}_{gain}$ to appear is:

$$\mathbf{t}_{time, delay\ of\ the\ gain} = \mathbf{m} \mathbf{V}_{gain} / \mathbf{F}_{force} ;$$

this means that the time delay for the $\mathbf{V}_{gain}$ to appears becomes smaller the higher the force is!

Then the product of force and time is the important impulse:

$$\mathbf{p}_{impulse} = \mathbf{Ft} = \mathbf{m} \mathbf{V}_{gain}$$

In exchange for this very simple proportional F=ma dynamic concept, we have to accept the far more difficult concept of the accelerated displacement section **length s** (h) measurement has now been **extracted / deleted** from the **Vis Viva** concept and **banned / buried** into a geometric vector-kinematic surface area **S** of the motion curve contour plot; furthermore, also the average velocity magnitude $V_{average} = \mathbf{dx/dt}$ has also been buried. The accelerated motion length quantity is here the area under the Velocity curve line and above the time base line, a quantity only extractable with differential calculus considerations when the motion velocity curve is not a straight inclining / declining line. When **the motion velocity curve is** a straight line incline / decline then the force and acceleration in respect to the time progression are constant then accelerated distance is vector-kinematical (s=h; a=g):

$$\mathbf{S}_{section, acceleration} = \frac{1}{2} \mathbf{a}_{cceleration} * \mathbf{t}^2_{time}.$$

This analysis of changes of velocity dV in respect to time dt is importantly called a **time domain analysis.**

This simplified correlation is so important that we assign to the **Force magnitude** causing a quantity of acceleration the designation of **Newton**, wherein one (1 N) Newton force is accelerating 1 Kg mass to 1 m/s velocity gain within 1 second delay. Newton viewed the time domain analysis as more logical and fundamental then the Vis Viva displacement domain because the force is also present with an accelerated displacement of zero, wherein the displacement domain returns an invalid infinite result for work passing a displacement distance of zero. In contrast, within machine mechanical constructs, displacement / acceleration distance of zero magnitude is also an illogical construct!

A valuable **advantage** of this time based domain analysis is that the force is also having a mean value **at half** the acceleration time duration if the acceleration is having a straight inclined or declined line progression in respect to the time progression, then the motion velocity-gain amplitude is proportional to the average impulse:

$$\textbf{F}_{\textbf{,average}} * \textbf{t}_{\textbf{time,delay}} = \textbf{mV}_{\textbf{gain}} ; \quad \textbf{I}_{\textbf{impulse,average}} = \textbf{P}_{\textbf{,momentum,gain}}$$

If the accelerated section distance **s,** (h), however, is not proportional to the mean value of a changing force in respect to time duration; the accelerated distance has to be extracted with a differential equation in the form of function of:

$$\textbf{S}_{\textbf{section,acceleration,distace}} = \tfrac{1}{2}\textbf{a}_{\textbf{average}}\textbf{t}^{2} + \textbf{v}_{\textbf{o}}\textbf{t}.$$

Newton's third law is based on this mean value force to the momentum gain proportionality. We can say that: within a **simple** opposing force motion (recoil) system, the momentum is conserved independent of the straight line Force slope, magnitude or direction of the force in respect to the time progression; this means the third law does not provide us with the actual equal momentum magnitudes, the third law only the pronounces the **equality** of the opposing momentum magnitude, because the accelerated distance "s" (h) is not included in this analysis. This principle will be fully explored for example #5 formula #8, #8.a.

The **greatest valuable** advantage is that we are able to diminish the velocity gain and the time duration in proportional parts into miniscule delta sizes **dv/dt** within the acceleration ratio (delta quotient) and we still arrive at the exact force magnitude for this tiny time section, a sort of snapshot analysis leading into calculus. This snapshot analysis technique is indispensable in the Inertial Propulsion proof analysis section and is also indispensable in electrodynamics, it is a fundamental principle.

Most significantly, Newton's time domain analysis returns a force magnitude **independent** of the actual **velocity magnitude**. The velocity magnitude could be zero or near the speed of light, it does not matter; only the incidental **velocity gain "dV"** in respect to the **time duration "dt",** the change in momentum, is analyzed; then, in calculus terms: $a_{,mean,value}=dV/dt$. In contrast, the "Vis Viva" is an analysis of the actual velocity quantity, the **greatness of velocity**, in respect to the accelerated distance. Newton's time domain analysis independence to the actual velocity magnitude aspect is an **advantage** and **a very significant disadvantage**; wherein the profound disadvantage is the time domains' isomorphic relation to the inertial mass motion directional kinetic energy flow leaving the first original root cause of the inertial mass motion unexplained, the isomorphic relation to the centrifugal acceleration, the friction work losses per displacement, the ability to do destructive work of the popular speed kills notion, the ability to bend objects, the inertial mass motion monetary energy cost and most importantly: the reciprocal motion speed magnitudes of the third law type motions remain unanswered. All these very important aspects of the two analysis systems must be kept in mind at all times and should be vigorous explained in our schools. The time based domain analysis can also be demonstrated with a declined notched board in Physics classes, the notches are spaced progressively further apart in a simplified 1,4,9,16,25 length multiplier progression wherein the 5th notch has a 25 times longer space then the first notch, a beautifully compelling progression of whole squared numbers of $1^2,2^2,3^2,4^2,5^2$. This provides a repetitive (isochronous, monotonous) click time interval. Both, the displacement and the time-domain analysis systems include a squared ingredient (parameter), the function of velocity squared or the quadratic growth of distance in respect to multiple repetitive occurring time intervals. Has anyone seen Galileo's and Newton's notched boards for side by side experimental comparison inside a Physics classroom? Why not side by side comparisons within classrooms is an important question to ask; a question already raised by the inventor of radar/radio communications principle H. Hertz in 1898. Throughout this present publication it is of premier importance to keep this measuring stick side by side relationship of Galileo's and Newton's notched boards in mind and let the students divide the boards length into Galileo's equal notches and Newton's progressive sequence notches.

Within his first Principia **1687** publication, Newton also presented his very important new invention of the **centripetal** acceleration solving the forces applying to an inertial mass moving in arc motions in opposite Force orientation to Huygens **19 year's** prior invention of the **centrifugal** force; they each have identical formula derivation and identical geometric vector-kinematic type pictorial derivation. This is because; the centrifugal and centripetal forces are a pair of inseparable forces in simultaneous opposing direction and these forces are derived in the displacement domain analysis of the arc radius displacement; because there is no obvious time delay d**t** of change in d**V** observable, only the inertial mass motion path direction is continuously changing! This means both, the centrifugal and the centripetal acceleration is a vector-kinematic displacement domain analysis, when the acceleration is multiplied by the mass it becomes the force, because F=ma. The centrifugal / centripetal forces are an inertial **force twin** impinging on a body in an arc motion, each does not exist independently without the other; furthermore, the centrifugal acceleration and the centripetal acceleration are proportional to the velocity squared in respect to the arc radius of gyration, a proportional-symmetric relation to Huygens Vis Viva Energetic principle, proportional to the work/Kinetic Energy principle while they are in an isomorphic relation to the impulse-momentum relation. Newton effectively shows that the acceleration can be either or in the time domain and in the displacement domain within a changing path, they are in both cases accelerations of an inertial mass. Newton was then applying the centripetal inertial mass motion analysis in a far reaching success to the planetary arc motion stability around the sun; allowing the re-formulation of Kepler's three geometric planetary laws into algebraic form. He successfully proved the validity of the invers law of planetary arc motion proposed by R. Hooke and the logical separation of weight from mass leading to the gravitational constant, and posthumous published a book in **1736** describing in detail how his analysis with "Fluxions" applies to inertial mass motion.

From his Principia writing and further statements it appears that Newton regarded the discovery of his first law of inertial mass motion leading to the centripetal acceleration, thereby leading to the prove of the Inverse law of stable planetary arc motions and the gravitational constant to be his most important work. Newton successfully argued that his centripetal force is preventing the earth from flying straight into the dark forbidding endless cosmos; this assigns the straight line inertial mass motion as a motion in absents of any force exertions based on Newton's first law. In contrast, Huygens didn't realise or couldn't believe that his centrifugal (away from the center) force is preventing, is continuously forcing, the earth away from falling into the Suns' immense gravity inferno, in a continuous tug of war of the inertial centrifugal force balancing the gravity pull action over a very large distance; but Huygens did indeed know, to his stated irate annoyance, that his **19 year prior** at the London Royal Society meeting reserved "accelerated displacement **length** based Vis Viva Energetic concept over the arc motion radius" is the singular first original root cause of both, the centrifugal, the centripetal and the planetary invers law force caused by the arc motion velocity **squared** in respect to the arc radius distance and the arc motion displacement. However, the suspicion in absents of proof, that the planetary arc motion was caused by a form of magnetic action over a large distance was already previously postulated (reserved) by Kepler, why did Huygens not agree in view of these historic events? It appears there was a simultaneous reciprocal ignorance of the Vis Viva by Newton and Huygens ignorance of Newton's inverse law proof! Today, Science still continues this tradition of reciprocal ignorance in view of Inertial Propulsion! Furthermore, Newton's ignorance of the heavy scientific lifting by Huygens / Leibniz inventing the kinematic Vis Viva principle started a well-established science tradition of ignorance based on derisions, a free for all self-serf scenarios. This Vis Viva analysis of inertial mass motion within **equal** repetitive **sections** of a circle is a fundamental displacement domain energy concept in congruence with Galileo equal distances notched board. Then planetary centrifugal force is:

$$\mathbf{F_{centrifugal}} = \mathbf{mass_{planet}} V^2_{speed,tangential} / \mathbf{r} = F_{sun\text{-}planet,attraction} = km_{sun,mass}\, m_{planet,mass} / \mathbf{r^2}$$

Accordingly, the planetary tangential Vis Viva magnitude is:

$$\text{Planet Vis Viva} = mass_{planet} V^2_{tangential} = \mathbf{k * mass_{sun} mass_{planet}} / \mathbf{r_{average}} \, ;$$

then, the planet arc motion average distance to the sun is:

#1.planet $\qquad \mathbf{r_{average,planet\text{-}sun}} = \mathbf{k * mass_{sun}\, mass_{planet}} / \mathbf{Vis\ Viva.}$

The constant "**k**" is the important gravitational constant invented by Newton; this means that the product of the Sun mass with the Planet mass divided by the sun to planet radius distance is proportional to the tangential Vis Viva of the planet. Then we can say: The Sun to planet average spacing is an inverse function of the planet tangential **Vis Viva Energy potency**; this indicates the VisViva is an important **fundamental principle in physics**. Furthermore, we can say: all Planets appear to have a common potential arc motion kinetic energy origin from outer space in relation to the sun center of mass; this correlation traverses the planet spacing by the coagulation theory; this is because, the planet orbital Kinetic energy is induced by torque and this torque cannot come from nothing. The primeval dust must have originated from an uneven clumped, in-homogeneous particle distribution to cause the planet angular orbital kinetic energy.

The previously reserved Huygens / Leibniz energy Vis Viva Energy potency concept is, of course, is not part of Newton's impulse momentum based Principia, the Principia therefore misses the above important Vis Viva correlation of the planets orbit radial positions based on the invers law and the centrifugal law. Then, during rotating pendulum experiments having simultaneous rotary centrifugal and straight line inertial momentum coupled reflections (translations) applying to a clock's escapement mechanism mass, Newton encountered also behaviors only solvable with force passing a radial distance displacement domain analysis, Huygens had described in his publications 19 years prior. In his first Principia Book published 1687, Newton performed a great leap of human intelligence and honesty, he sensed a more complex system within these centrifugal combined with straight line momentum motions, calling theses investigation too numerous and tedious for final analysis. To keep his Principia uncluttered and to avoid using or referencing Huygens Vis Viva energy principle, he made a very smart move by separating straight line displacement mass motion from the troublesome combined centrifugal to straight line motion coupled (translated) pendulum experiments and apparently allowed future scientists to develop better Physics tools describe these systems. The answer of the present day scientific understanding explaining Newton's omission is: Newton's three laws of motion apply only to point mass sizes **without complex** Cartesian plane motions, an unknown science at the time of Newton. Evidently, we have an unfinished theorem of a great scientist, like Fermat! Fermat ran out of paper and Newton ran out of time-allotment and patience. Newton seemed to cast these pendulum experiments not only off into an uncharted area, but he cast the subject off Science limit by an inconsistent all-encompassing pronouncement. Newton postulated his third law of mass motion by arguing that there is **always** an equal and opposing reaction to **<u>any</u>** mass motion action. The **<u>ALWAYS</u>** argument appears to include also Huygens combined straight line displacement to rotational motion reflections against pendulums employed within clocks. This **always** statement is un-provable because Newton himself stated near infinite possible inter correlation reiteration of the three possible motion directions of one mass unit, correlated with infinite velocity progressions, internal (Leibniz type) friction losses, then correlated onto multiple mass units. Wherein the motion directions comprising: The axial rotation, the tumbling head over heels motion and the overall forward motion. Newton writes about his combined rotational pendulum motions reflected against a straight line displacement inertial mass motion, presenting here his own words from the Principia reprint "On shoulder of Giants by S. Hawking; Newton speaks here to us: **"….. From such kind of reflection also sometimes arise the circular motion of bodies about their own centres… .. …But these reflections <u>I will not consider</u> in what follows and it would be too tedious to present every and all examples of these combined motion reflections"**.

Here, Newton effectively quarantined a larger segment of science from our knowledge and he never rescinded this **vague "Third Law exception"** in any of his further Principia's. The sad and disturbing fact is that science has filed Laplace's unfinished omission as a to-do item, only recently resolved, while Newton's omission was filed as a **NOGO** area because of an old, excluding modern dynamic breaking technology, conservation of momentum conflict assignment onto all possible inertial mass motion mechanical configurations; this is undoing Newton's wise and honest: "But these reflections **I will not consider**......." admonition! We ought to assume that a great accomplished scientist of Newton's importance is presenting these inter-correlations for a very good reason. However, did he realise that he set forth a long lasting culture of exclusion, ignorance and preferences contrary to a full wholesome scientific investigation. Then we must reasonably conclude from these exclusion-inclusion statements that Newton already presented us in his first "Principia" the answer of how Inertial Propulsion can work; it is Newton's cautious admonition how IP works: **Inertial Propulsion works with independent rotational mass motion reflected (translated) for every individual rotational displacement position onto a straight line inertial mass motion**; this is a Vis Viva analysis in respect equal accelerated displacements intervals in symmetry with the centrifugal force derivation; it is not an analysis in respect to equal time intervals because the time intervals for every angular displacement section are in most cases, like the pendulum, the scotch yoke, the combustion engine and etc.. etc.., variable. This is most famously the physics of the pendulum discussed later. In retrospect, in view of this somewhat obscure Third Law exclusion-omission, it can be assumed that Newton already had performed an underlying un-published experiment indicating Inertial Propulsion is probably possible. Newton already **limits** here the universal **reach** of momentum conservation within machine constructs; this is a most fundamental principle based on the division of the circle into angular displacement parts for rotational displacements reflected (translated) onto a straight line motion. Obviously, we must divide the circle in correlation of the straight line reflected (translated, coupled) motion length within machine constructs: **This is because, the cyclic events within the angular displacements of a circle reflected (projected, translated or coupled) onto a straight line back and forth motion <u>repeat only in even angular displacement events</u>**. This division principle places the rotational motion reflected onto a straight line motion firmly into Huygens / Leibniz Vis Viva displacement domain analysis and places it in symmetry with the centrifugal forces, friction forces, chemical-gas forces and permanent deforming forces of objects. This universal angular distance division principle is also present in our electrical and thermal-chemical-energy motors wherein the angular motivating impulse is in angular displacement synchronisation with the rotating inertial mass of the motor rotor or in angular displacement synchronisation with the motor crankshaft with ever repeating angular displacement length, it is the very rock bottom

<u>**far reaching principle of physics;**</u>

it is in all cases forces passing an angular displacement of repeating distances having each variable time durations. We also find this division principle of the circle in our calendar derivation, the Physics of our Earth planetary motion. The Earth angular cyclic motion around the sun are in precisely repeating 4 angular displacement distance quadrants of Solstices and Equinoxes in an endless fractional relation to the planet Earth rotation time / days count requiring progressive leap years to account for these fractions; this principle was known to the ancient Stonehenge builders, but blatantly disregarded for Roman imperial pan-European political reasons by the builders of the Julian Calendar paid for by the Roman military / political leader "Cesar" in 46 BC. What compelled Christian followers of the **Prince of Peace** to forcibly impose this difficult numerical calendar, commissioned by a belligerent conquering enslaving pilfering military leader, onto the central European barbarians treasuring their Physical Calendar festivity dates based on the culture of continually observing the heavens for Solstices and Equinoxes? From this presentation we can say inertial Propulsion is not necessarily contrary to Newtonian Physics because of the third law exception based on tediousness leads to an inclusion based on curiosity. Blanket rejection of IP without a careful analysis including **all** aspect of modern physics, including Newton's third law exception, the third law in the feedback energy form in combination with the dynamic breaking principle; this would be contrary to Newton's vigorous scientific discipline and human honesty. At this point in science, a blanket rejection of IP must be viewed as a rear view on the meagre quantity outcomes of past IP experiments and an unreasonable exclusion of half of our science knowledge. Within this publication a stream of examples experiments are presented to prove positive the IP principle to exhaustion by including Newton's exceptions to his third law analysis.

The return of Newton's stated exception to the Third Law

The question / suspicion whether or not a self-contained third law type force impulse can exist within an isolated system of a vehicle was raised again when clockmakers attempted to build clocks capable of sustaining the local time of the port of departure for longitude navigation. Here again, we have Huygens' rotational pendulum mass motion with straight line displacement reflection being employed within these clocks.

Huygens was heavily involved, from the very beginning of research, in finding the perfect clock for ship navigation; he identified a more complex problem here because of his reported mysterious exact time synchronisation of two pendulum clocks standing in physical contact side by side.

Clockmakers were confronted by an intriguing problem. It seems, no matter how ingenious such clocks were devised they either advanced or retarded when placed on ships in comparison to the port of departure local time. This of course means that clocks gained energy or depleted energy over time, while clocks are designed to deliver very exact equal energy portions over very long time duration. It took more than 100 years of frenzied scientific work and the incentive prize of a king's ransom to solve this scientific challenge.

The final successful idea by Jon Harrison was the determination that the complex motion of the ships was causing changes in the clock timely energy distribution magnitudes involving the kinematics of two motion dimensions the centrifugal and the straight line forces on inertial mass motions. Here we have the return to Newton's exclusion from his Third Law in the form of **a perfect chronometer clock** exist when **all straight line motion** reflected (translated) from rotational motion **are extracted**; any reflection (any translation) from rotational motion coupled onto a straight line motion will interfere reciprocally with the clock and the ship. Is the chronometer clock problem disappearing for modern atomic clocks wherein electrons have a miniscule mass of $9*10^{-31}$Kg? According to Einstein: atomic clocks must be perfect chronometer clocks at the speed of light! This chronometer clock principle is the theme of the endearing film "Longitude". In this true story film, the clockmaker and carpenter Jon Harrison, determined that his experimental test clock was affected by a certain motion of the ship that his clock creation was tested on. He was able to extrapolate the time delay of the clock to the changes in initial potential energy conditions of the clock pendulum swings caused by the ship motion impinging on the pendulum motions. Harrison determined the amounts of time delays per pendulum swings thereby saving the ship from a navigational disaster. This Longitudinal navigation film story is documenting a brilliant performance of human intelligence by the carpenter Jon Harrison. This appearance of fine human intelligence in unusual places is a repeated phenomenon where a Diplomat designs the first four math function hand cranked mechanical calculator, tailors build the first human piloted hot air balloon, a mining engineer builds the first hang glider, bicycle mechanics build the first powered flying machine piloted by humans and a patent clerk is the solver of science mysteries. How can we explain this ship mounted clock phenomenon with Newton's Third Law in the time domain of ALWAYS equal reaction to an impulse action? How can an action of the isolated system of a ship react onto the kinetic energy of a clock contained on the same ship without direct mechanical transmission traction simply by "sympathy" of the rotating and oscillating inertial mass motion, how can it be caused by fictitious inertial forces? Since the ship to clock energy transfer relationship is a documented reality, then it can be argued correctly that because of the reversibility and scalability of physics principles, energy and impulse must be continuously transferable from very large clocks mounted within vehicles in a reversed process; this clock to ship transfer principle has been proven by the Roy Thornson's US patent No: 4631971 employing the centrifugal Formula #1.10; this experiment is on the "youtube internet:

www.**youtube**.com/watch?v=Bx4LT3GZjlY.

However, the political appointed gate keepers of scientific truth try to **dismiss** such irritating Inertial Propulsion phenomena as caused by reiteration / reverberations / stick-on / friction against the surface of the earth without delivering a formal physics description / quantifiable proofs of these hidden actions simply by the strength of their assigned authority. In response to the stick-on argument, it will be experimentally proven with **pendulum tests** that **all** presented **four** mechanical proven performances **I**nertial **P**ropulsion constructs presented here **do not work by the principle of friction to the ground;** their quantifiable internal inertial self-contained impulse performance (potency), however, **is reduced by internal / external friction and energy absorption!** According, applicable to the presented mechanical designs, the stick-on friction argument must be redefined-corrected to be applicable to purely straight line uniform motions with small friction compliant acceleration differentials. This principle of the separation of friction from inertial mass motion was already presented by Galileo with his notched boards! If we need the surface of the earth as a reference source to motivate a vehicle with a self-contained impulse, why is it not possible to use a second clock delivering an identical directed impulse magnitude but in a mutually opposing rotational mass motion direction mimicking the reference source? Yes, this publication seeks to present that such a system of tandem mechanical oscillators have a unidirectional self-contained impulse capability **generating its own reference source;** this principle was postulated by Prof. Eric Lathwait. This publication's aim is then to provide an answer, in formal science terms, to what these reiterations / reverberations / stick-on / frictions are which motivate vehicles without traction of wheels or without expulsions of masses. Accordingly, in view of the ship chronometer reality without any further ado, we must already concede that inertial propulsion must be possible and patents claiming such capability must be carefully examined for individual validity.

The question still remains: "What thrust magnitudes are possible within what type of mechanism, what kind of math returns the real quantity of thrust?".

A comparison of the two Domain Analysis Systems

At this point, having viewed the basic principles and history of mass motion analysis, it is important to compare the underlying physics analysis views pertaining to the displacement domain analysis and the time domain analysis. What is the underlying physics analysis of the same physics phenomena of inertial mass motion explained in an indisputable practical format? The displacement domain analysis is telling us that the nature of inertial mass reluctance requires a progressively larger force exerted per **equal** (repetitious) distance intervals to increase the mass motion **velocity** magnitude in **equal** increments. This is because an increase of mass motion velocity instills into the inertial mass a larger ability to do work; the Vis Viva potency is **depending** on the previous speed history of the mass motion velocity. The Force is required to perform work over the distance **h** to change the "Vis Viva" potency magnitude; accordingly, we divide both side of the VisViva formula by **h** and arrive at the Force due to gravity is:

$$\mathbf{F}_{force, \, gravity} = \mathbf{mg}:$$

$$\mathbf{E}_{energy}/\mathbf{h}_{,height} = \mathbf{m}_{,mass} * \mathbf{g}_{,gravitational,acceleration} = \mathbf{F}_{orce} = mV^2/2h_{,height}$$

The Vis Viva / Energy can be vector-kinematical generalised as a Force in respect to the accelerated displacement distance analysis arriving at formula #1:

$$\#1) \; \mathbf{Force}_{,average,value,}N = a * mass = \frac{mass \, (V^2_{new,speed} - V^2_{previous,speed})}{2S_{section,acceleration,distace}}$$

Ref: Gieck Formulas: Kinematic section L1 to L10

$V_{previous,speed}$ is also presented as V_o (Velocity Origin) in many publications.

1a: $VisViva = \mathbf{E}_{energy} = mV_{gain}V_{average}$; $V_{average} = s/t$; $t = mV_{gain}s/E_{energy}$

From this Force in respect to the displacement formula we can extrapolate that the **early displacement POSITION**, within a long accelerated distance motion quantity, where the maximum gain in speed per **equal spaced** displacement is occurring will significantly influence the sum of the FORCE mean value magnitudes in respect to the distance and will significantly influence the average speed magnitude of the motion in respect to the total motion distance. The initial potential and kinetic energy sums and the impulse

26

sums remain zero (0) within an isolated system. This is the Lagrangian, Hamiltonian and Third law principle.

It is also very important to note nine (9) important logical points:

Point **#1):** Formula #1 applies <u>only</u> to a motion exerted from a fixed point, like the earth gravitational pull or a gunshot without recoil, etc. For derivation of motions with recoil please refer to Example #5, Formula **#8!**

Point **#2):** The number **2** as divisor, in this formula, tells us that formula#1 applies to a displacement section having a mean value force at half distance of the total accelerated distance and the acceleration is a straight horizontal line, straight incline or decline in respect to the incremental distance; this means the force is steady, increases or decreases at a constant amount for every uniform measure of distance interval; this universal concept is most commonly and easily presented with a compression spring or a stretched elastic.

Point **#3):** The time duration for each displacement section is the section distance divided by the average speed:

$$t = s_{,section,distance} / V_{average}; \text{ wherein: } V_{average} = (V_{orgin} + V_{new})/2 = s/t$$

#1.a1) $$t = m V_{gain} S_{section, distance} / E_{VisViva}$$

This means that the Inertial Propulsion machine herein presented is having variable angular displacement time durations for each ¼ angular turn displacement; because the energy magnitude delivered into and withdrawn from the system is in relation to the repeating ¼ angular turn displacement distance, a formula #1 displacement domain system wherein the section time duration is an inverse function to the energy magnitude delivered; this means: more energy causes a shorter time duration in a diminishing return progression. The diminishing returns progression means that the energy availability will run out empty before the time duration reaches zero.

Point **#4):** Using a computer program to arrive at smaller division of the distances and smaller increment of average forces does not change the underlying Physics principles; it only improves the accuracy of the calculation.

To visually clarify this displacement domain concept a graph of a steady changing Force from $4N_{Newton}$ to $0.0 \ N_{Newton}$ in respect to 10 repetitive displacement distance increments to a total of 1 meter is presented with the graph on the next page:

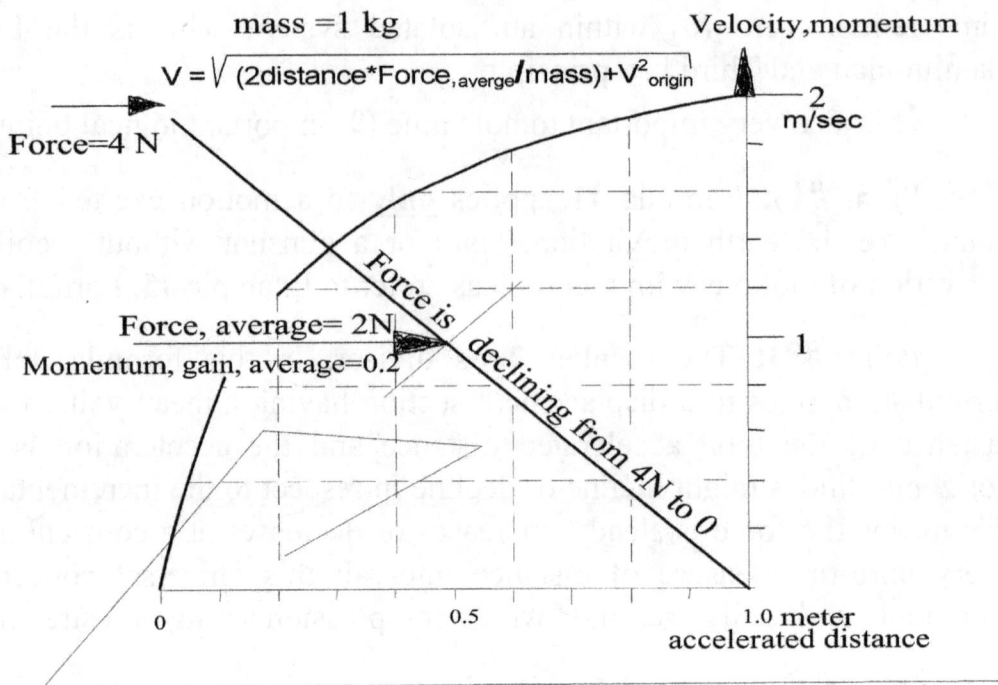

mass =1 kg

$$V = \sqrt{(2distance*Force_{averge}/mass)+V^2_{origin}}$$

Velocity,momentum

Force=4 N

Force is declining from 4N to 0

2 m/sec

Force, average= 2N
Momentum, gain, average=0.2

1

0   0.5   1.0 meter
accelerated distance

Kinetic energy=Force,average*distance=velocity,gain*velocity average

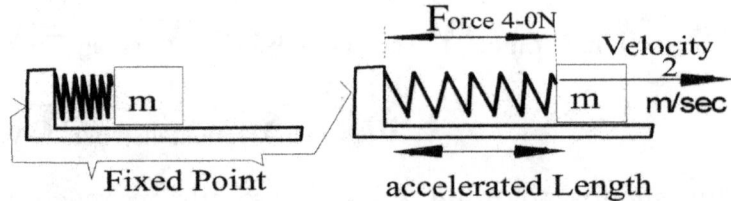

Force 4-0N

Velocity
2 m/sec

m

Fixed Point   accelerated Length

:

We then say: The acceleration applying to formula #1 is constant or is changing at a constant rate while passing over equal incremental distances. It will be shown with many examples that the 2 divisor is **not** universally retained in all displacement domain mechanical constructs, in many cases it cancels out.

Point #**5**): The inertial mass motion parameter magnitude correlations have an exhaustively experimentally verified validity by the G. Carioles and many other scientists. With today's Science Knowledge this Ke parameter validity is a universally valid principle in Physics. The most important, simple, original and famous experimental verification is Galileo`s inclined notched board with equal spaced notches.

Point #**6**: the difference of new squared speed minus the old squared speed is a gain in inertial mass motion kinetic **energy** and should be logically viewed as the gain in velocity squared in a geometric vector kinematic analysis.

Point #**7**: The energetic Force is empowered to follow the acceleration distance because the velocity squared is proportional to the original potential accelerating mechanical work / energy in form of the **product of force and distance**, and because the distance is the accelerated motion degree of freedom.

After the inertial mass motion passed through the accelerated distance Newton's first law applies: An object in motion stays in motion with the same speed, the same Kinetic energy and in the same direction unless acted upon by an unbalanced force.

Point **#8)** Formula **#1a, #1a1** must be viewed as the base fundamental and **the first argument for IP;** this **is** because, within a rotational to straight line coupled (translated, or **Newtons's reflected**) motion mechanisms the cyclic repeating straight line alternating opposing back and forth momentum amplitude differences are $(+mV_{gain}-mV_{loss})=0$, and each $V_{origin}=0$), and the associated angular displacements distances are sequentially repeating; this **appears as** a Third law limiting scenario; however the Author presents that the angular centrifugal / centripetal Force must also be considered a direct consequence of the Force in respect to the displacement analysis, also called the Work / Energy theorem wherein the kinetic energy of a rotating vector mass is considered de-accelerated in **a straight line $F_{force,centrifugal}=-mV^2/r$,** and then re-accelerated and transferred into a 90° turn angle $F_{force,centrifugal}=mV^2/r$; the 2 divisor cancels out reciprocally and the Kinetic energy of the rotating mass is conserved. Accordingly, the centrifugal / centripetal force is:

$$F_{force,centrifugal}=mV^2/r,$$

is congruent with the Vis Viva because **s=r** and 2 cancels out! The work done / energy exerted and energy recovered by each ¼ turn is the centrifugal force in

energy balance form: $F_{force,centrifugal}=mV^2/r$;

this means if the velocity V is increased by Leibniz calculus of changes dV amount, then both, the Force and the radius each increase relatively a proportional dF, dr amount and still maintain the equilibrium of energy potency! The first mechanical measurement and mechanical government control of an engine speed was accomplished by James Watt with the simultaneous measurement of the changing of the centrifugal pull force **dF** and the changing of the radius **dr**

#1.10) $(F_o+dF)(r_o+dr)=m(V_o+dV)*(V_o+dV)$

in response to a changing velocity **dV** causing a measurable work in the lifting of the orbital inertial mass motion distance **dr** on levers or the extension of a spring in response to the changing dF_{force}; accordingly, mechanical speed measurement was firstly in relation to a Ke / work energy potency measurement. Therefore, in view of James Watt measuring device of velocity with the centrifugal force formula derivation and the related #1.10 Lagrangian equilibrium of energy / work derivation the notion of the fictitious centrifugal force is **redundant**; it ought to be dispatched from our science arguments and replaced by the Lagrangian energy balance!

Ref. www.physics.princeton.edu/~mcdonald/examples/governor.pdf

$$F_{force}=mV^2/2s; \ s=r, \text{ the } \mathbf{2} \text{ cancels out}$$

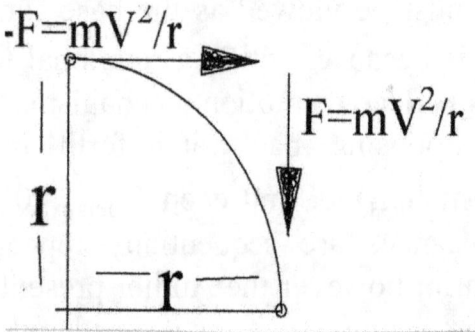

However, very importantly, the cyclic $V_{average}$ for each starting and stopping motion can be forced variable with the Authors patented combined centrifugal / straight line motion IP mechanisms based on formula **#1a**, wherein we arrive at variable energy flow and a congruent variable centrifugal forces within the two equal opposing motions displacements derived from angular rotational reflections (translation) applying to an isolated system; wherein, again, we have to remember that the kinetic energy and the centrifugal forces are in direct proportional relation. This is called a **F**requency **M**odulated (**FM**) cyclic inertial motion, a complex cyclic motion when the cyclic motion is a combined tethered centrifugal motion, the straight line inertial mass motion in the displacement domain and the centrifugal forces have dominance. This is Newton's exception to his third law, it can be defined by formula #1.a.2 wherein $V_o=0$:

#1.a.2) $\mathbf{I}\text{mpulse}= \mathbf{E}_{energy}/\mathbf{V}_{average}$; because: $mdVt=mV^2/2(s/t)$;

Formula **#1.a.2** means that the larger the first original root cause energy and the smaller the average motion velocity $V_{average}$ the higher is the impulse!

Within the following graph in the next page the "**a***distance" product term is presented as a shaded area, when multiplied by the mass magnitude it is logical a volume; this volume is proportional to the volume of Oil, Coal or Gas for the same amount of energy, when multiplied by its volume to energy conversion Constant. Accordingly, the $\mathbf{V}_{gain}*\mathbf{V}_{average}$ =Vis Viva derivation is easily visible within the graph, it is visible that it is **not** related to ½ **mass**, it is related to the

Speed$_{average}$=s/t=distance/time

of the accelerated displacement section; this is so often misstated in our schools and ought to be vigorously corrected.

Point **#9:**.Not everyone instantly visualizes this algebraic principle of differences in squares within formula#1 and how the squaring of velocities relates to Newtonian mechanics. Accordingly, for illustrating **the power of the Vis Viva** function in vector kinematic proof form we draw next a compelling simple visual fundamental graph picture wherein the velocity origin magnitude plus the velocity gain is presented as a Newtonian total line length vector, this is placed side by side with the Huygeanian square of the velocity origin inside the velocity origin **+** velocity gain square. Here we have the displacement domain analysis on the right and the same velocity gain in the time domain on the left. The product of velocity gain and mass is momentum; the product of velocity gain, velocity average is the **a**$_{acceleration}$*distance and **is the shaded area within the graph** when multiplied by mass it is energy; accordingly energy is the product of two velocities and mass: Formula**:**

#1a) $\mathbf{E}_{energique}\mathbf{=mV_{gain}V_{average}=mF_{force}S_{section,distance}}$ **=Vis Viva**

This is congruent with the **Reference**: Gieck Formal Book section **L5.**

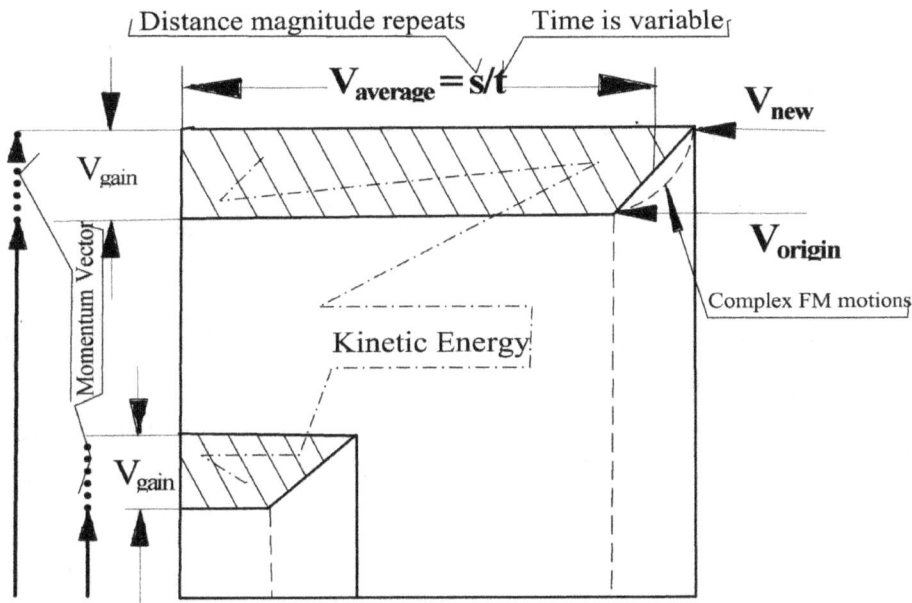

From the above impulse-energy graph it is obvious that the Vis Viva and energy is growing in quadratic progression of the velocity (greatness, amplitude) magnitude, while momentum grows proportionally with the velocity gain magnitudes. Here we have achieved the H. Hertz admonition that we must first correlate impulse with energy to achieve technological advancements.

Furthermore, for a uniform motion the average speed is the speed sum divided by **2**.

For a non-uniform force the average speed is the sum of all individual average speeds then divided by the number of sections **n**:

#1.1) Force,avera.=(force,sect.1*dista.+force,sec.2*di.+force,sec.3*d..)/Dista.total

#1.2) average speed =average,sect.1+average,sect.2+average,sect.3.../n

#1.3)**Impulse**sum=force sect.**1***(distance/speed,average,sect.1)+force,sect.**2***(d......

From this displacement analysis formula a high resolution, accuracy and certainty of energy, average force and impulse can be obtained from any type of motion depending only on the number of sections used. Furthermore, from here, the isomorphic behavior of inertial mass motion speed in respect to energy can also be extrapolated. However, Inertial Propulsion is performed with a combined rotational and straight line displacement motion reflection in a non-uniform rotational motion progression having a mutual reciprocal exertion between two inertial mass bodies without an external fixed point of exertion (Newton's too numerous and tedious experiments) wherein the 2 divisor is applicable to both, the changing **angular** motion, the changing **straight line** displacement motion. We must apply the consideration of potential energy distribution between two bodies, a triple averaging procedure. Accordingly, we can postulate that the Ke, work theorem of the displacement domain is a very powerful analysis tool including **all 6 possible inertial mass motion parameters, at once, in one single formula package:** 1)**Force,** 2)**acceleration,** 3)**accelerated distance,** 4)**time,** 5)**velocity**gain and 6)**velocity**average to solve many complex inertial mass motion problems!

In contrast, the **time domain** analysis is telling us that a sum amount of impulse, the summed product of force and time duration, will impart an increase of proportional amount of inertial mass motion velocity independent of the time position, independent of displacement length, independent of the degree of accelerated freedom, previous motion history or velocity magnitude (greatness) pondering, wherein the **force is assumed to be empowered to follow** the inertial mass speed gain if no opposing force is present and the force is exerted from a fixed point; there is **no first original root cause** presented here to proof what is causing the force to follow the acceleration! Then the **mean value** Force, a synthetic mathematical-dynamic concept is:

Formula #2) $F_{force,mean,value,N} = mass * S_{peed,gain,straight,line} / time$

Wherein the speed gain is:

#2.1) $Speed_{,gain} = (V_{,new,speed} - V_{,previous,speed})$.

The mass multiplied by the speed gain is the change in momentum=d(mV).

The time duration is: Time=Distance/$V_{average}$

For $V_{orgin}=0$; $T_{ime}=(2mass*distance/F_{orce})^{1/2}$

$V_{mean,value}= s/t = Distance/time = V_{average} = (V_{new}+V_{previous})/2$

This publication uses $V_{previous}$ and V_0= velocity origin exchangeable.

Seven Important points:

Point **#1)** The Third Law of equal reaction to an action is derived from the **difference** in speeds within formula #2 independent of the speed magnitudes, therefore, it analyses the reciprocal relation of impulses in relation to their resulting equal momentums reaction; accordingly, it is **not** delivering the **actual magnitudes** of the reciprocal **momentums** it only postulates the equality of the reciprocal applied impulse. For calculation of the **motion actual magnitudes needs formula #1**. This is proven with example #5, Formula #8.

Point **#2)** The use of the difference of speed magnitudes and average speed magnitudes provides compatibility with formula #1, it provides the kinematic **bridge** into formula #1.

Point **#3)** Formula#2 **excludes** the direct concept of average speed; it also **excludes** the **direct** concept of degree of freedom of accelerated motion in form of the accelerated /de-accelerated distance "$S_{,section,distace}$"; to obtain the accelerated distance a separate formula and calculations is needed for each consecutive time sections having a change in force / acceleration magnitude. The accelerated distance is: Formula **#2c**: $S_{,section,distance}=\frac{1}{2}at^2+V_0t$.

Point **#4)** The halve distance, **s/2,** of a constant accelerated motion wherein the **velocity curve** is a straight inclining / declining line, when starting from $V_{origin}=0$, occurs at $t_{total}/2^{1/2}=t_{total}/1.414$ independent of the acceleration magnitude. This is a purely geometric / kinematic argument because the distance is the area under the velocity curve.

Point **#5)** When we attempt to apply formula #2 to rotational dynamics within machines, piston Engines, pneumatics, cam motions, oblique pendulums,

Trebuchets or IP machines, where the straight line displacements are Newton's reflections (translation) in relation to an ever repeating angular distance division of the circle and where energy transactions occur, it becomes only solvable with formula#**1a,** or with complicated multiple derivative of time (multiple formula#2); because within machines the displacements magnitudes are rigidly cyclically repetitive and the time duration is variable. Formula #2,#2c delivers a displacement area under the velocity curve and is invariable an endless fractions of displacement distance; accordingly, **only formula #1,#1a** can solve equal repetitive displacements efficiently. This is the reason for Newton's exception to the third law related to formula#2.

Point **#6)** When we view the time domain analysis with Newton angular reflection (translation) of straight line motions in mind, we cannot visualise that such a rotational motion linked with straight line motion could be frequency modulated and Newton proposed, in his post Humus book on "Fluxions", a derivation methods of the third time derivative of the change of momentum to solve this shortcoming; while in the displacement domain such a possibility is clearly logical possible, because variable time is a parameter member of formula#**1a**. Frequency modulation means that the angular velocity and therefore the centripetal acceleration are variable while the straight line **velocity amplitudes** are constantly repeating; this is again a reason why Newton separated this possibility from his analysis in the first Principa book.

Point **#7)** The computer program herein presented to calculate finer increments of change of momentum in respect to the time duration does not change the physics of this concept, it only changes the accuracy of this correlation. Computer programs, however, are a great help changing from the displacement domain into time domain analysis wherein the acceleration magnitude and the average velocity are the common factors of both analysis systems.

Point **#8)** The time duration of the accelerated motion and the time in the denominator of the formula #2 is not a degree of freedom of motion, it is invariable constantly repeating, accordingly frequency modulation cannot be easily visualised with formula #2. The accelerated freedom of distance is the **area under** the acceleration curve, a quantity in relation to V_{origin}, $V_{average}$, the acceleration magnitude and the time duration.

The change of displacement in respect to time is in the time domain a separate calculation; accordingly, formula #2 is a synthetic mathematical construct derived from formula #1 by removing the concept of average speed.

The mechanical construct to simulate formula #2 needs a clock timer operating an energy based accelerator (in reference to the car accelerator) energy valve to dispense the desired Force over the time duration=impulse magnitude from a store of the **first original root cause** potential energy in respect to the timer time progression; wherein the store of energy, is for example, a robust electrical energy reservoir, a gasoline tank reservoir, a pneumatic energy reservoir or a nuclear energy reservoir. The correlation of impulse to total stored potential energy is the formula #1a, #2a function. The extreme non-uniform speed progression for an IP device is also indicated in a dashed line.

Ref: Gieck Formulas: Kinematics L1 to L10

To present the important points **#1 to #8** in a graphic presentation a picture is provided on the next page:

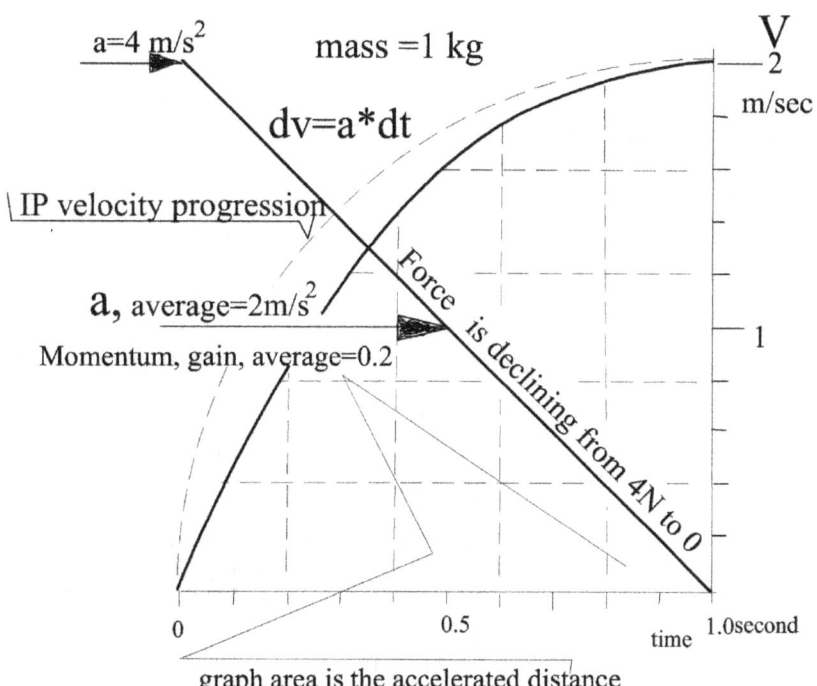

graph area is the accelerated distance

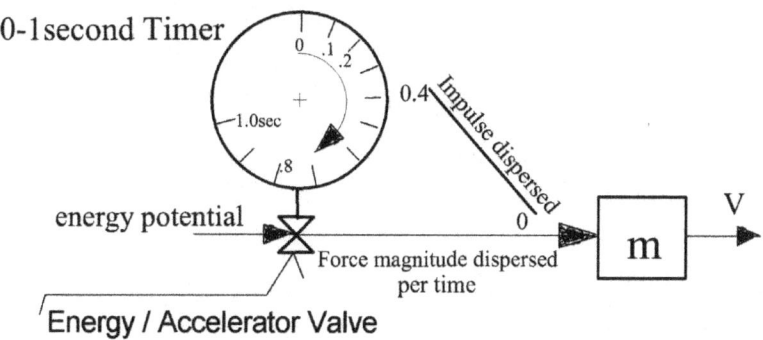

35

Within formula #2 we got rid of the Cariolis averaging factor in the denominator which applies to the constant acceleration still present in formula#1. Formula #2 applies to the mean value of force within a uniform motion profile; therefore no divisor of 2 is required. This aspect of the time domain analysis is fully explored in the rotational motion reflected (translated) onto straight line motion section.

Examination of the absolute speed limit of Inertial Propulsion

From the logic of formula #2 there is no logical reason for any speed limit in respect to the previous speed, or the speed gain or speed magnitude, because speed gain is a **difference of speeds** wherein the speed magnitude (amplitude or greatness) disappears, the speed magnitude is simply ignored! Accordingly, force is the change in momentum in respect to time.

Newton himself, when asked to define the absolute speed limit of inertial mass motion, Newton answered:

"It is infinite".

Accordingly, we must initially conclude that, very likely, machines propelling, "accelerating", past the speed of light c might be possible; because Newton already provided us with an exception to his third law! However, within this speed limit of inertial mass question, it can be argued that if the previous speed magnitude amplitude V_o starting the speed gain is great enough, then one second displacement time duration will cover the whole distance of the universe; for a reasonable accelerated distance exerted from a fixed point by a vehicular mechanism, the impulse time duration will become hyper short that the vehicular inertial mass will be unable to respond to such an extreme miniscule impulse duration. This is the Huygensian relativity. The Huygensian relativity needs formula #1 to reason it; but it ignores the **divisibility limit** and granularity principle within Physics, a universal principle also found in Newton's corpuscular light theory, Electrodynamics, Thermal, Nuclear, Quantum physics, and the oldest original Atom theory of antiquity, the oldest rock bottom uncontested physics principle ever postulated. We now illustrate the two inertial mass motion analysis systems side by side in practical usage terms to analyse the maximum speed question. In the most concise terms, the displacement domain analysis is the analysis of the actual velocity **magnitudes** (the velocity greatness, amplitude) in relation to the accelerated displacement section **"s"**; while the time domain analysis is an analysis of **velocity differences** in respect to the time duration. We will look at the operation of an ideal race horse having its maximum speed gain at the race finish line and having weightless-mass-less-frictionless legs; also referencing the horse race graph picture on page 52 and the reciprocal impulse machine problem #5, formula #8, #8.a:

For the displacement domain we can say: The horse forages on oats which has an equivalent of energy printed on the serial box in Kcal which contains a proportional equivalent of force multiplied by displacement distance Kgfm = work, 1 kcal or 0.0023 kgfm. This means there isn't much energy in terms of kgfm in a box of oats. The horse must move its legs for every uniform measure of distance sections, displacing its body-mass with a force which is depending on the previous speed of its body-mass according to the work-energy theorem:

$$\textbf{\#1b) Ke=Force * distance, Nm} = \frac{\textbf{mass} * (_{\textbf{new,}} \textbf{speed}^2 -_{\textbf{previous,}} \textbf{speed}^2)}{\textbf{2}}$$

Formula **#1b** means that:

The faster the horse runs the higher is the required force per equal spaced distances of the steps to obtain a repeating magnitude of velocity gain, in a quadratic escalating progression. We easily conclude that the speed of the horse is limited by the magnitude of the force it can deliver over the uniform measure of distance. Here again the "Vis Viva Principle" is at work.

The quantity feed of oats the horse has previously received is its stored potential energy and the ability to deliver this energy to its legs per time is the Horse Power. **T**hen Horse Power is:

#1.4:
$$\textbf{P}_{\textbf{power,}} \text{HP} = \textbf{F}_{\textbf{force,average}} * \textbf{V}_{\textbf{speed,average}} ;$$
$$\textbf{V}_{\textbf{speed,average}} = \text{P}/\textbf{F}_{\textbf{force,average}}$$

Accordingly, within inertial mass motion exerted from a fixed point (a **non type** #5 formula #8 machine), energy expended reaches infinity well before mass motion speed magnitude reaches infinity and also, the energy expended reaches infinity before the **impulse time** duration reaches **zero**. This is the root of the Galileo-Leibniz-Huygens-Hertz Kinematic Relativity Principle; a logical energy conservation constraint argument. When we experimentally determine that the speed of light=**c** is the extreme speed limit of an inertial mass motion when exerted from a fixed point, then we have Einstein's Relativity principle. Einstein's relativity principles are experimentally derived from a mass motion exertion from an unmovable-**fixed point** with drive energies far exceeding the required energy magnitude determined by formula#1, wherein the absolute speed limit **c** has been determined to be the speed limit of an

electron mass$\approx 9*10^{-31}$Kg

accelerated up to the speed of light by an unmovable (static) **hyper** high-voltage cathode-ray tube electric field; wherein the kinetic energy of the electron reaches **Ke=80*10^{-15}**Nm(fempto Newton meter). The cathode-ray tube is also called an electron gun. This principle is also encountered with the large Hadron accelerator, wherein many electron guns are arranged in a circle from muzzle to breach, each gun barrel section provides a straight line vector impulse and also provides a slight centripetal impulse for bending the electron beam into a circle. The speed measurement of the electron beam is the voltage magnitude induced within a wire loop (today many more sensors types are available) by the cycle frequency of electron beam sections rotating within the gun circle. The $\mathbf{S}$pecial $\mathbf{R}$elativity $\mathbf{T}$heory, **SRT**, is accordingly based on the physics of a hyper miniscule mass in an arc motion receiving hyper high motivating impulses near the speed of light from consecutive **fixed points over repeating distances**; wherein the impulse durations for each accelerating one meter gun sections are hyper short, they are in the nano (10^{-9}) second range. This experiment found that the acceleration speed gain near the speed of **c** is un-measurably small in comparison to the extreme force applied. Accordingly, the absence of speed gain **appears in**congruent with F=ma. To fix this F=ma apparent incongruence, **SRT** uses a "positive kinematic feedback term":

#1.e $\qquad\qquad\qquad$ $\mathbf{(1\text{-}V^2/c^2)}$

borrowed from electrodynamic radiation and from the Lorentz factor within light illuminations of relative motions near **c** without presenting a visual / kinematic feedback picture proof of **what** is feeding into **where** within the actual inertial mass motion under analysis! How is feedback fitting into the dynamics / kinematics of inertial mass motion exerted from an inertial frame at the speed of **c** in a logical **comprehensible kinematic** feedback flow chart? Legions of scholars agree that Einstein's SRT electro dynamics of inertial mass motion comparison, at **c,** is a poorly founded derivation; this is because in macro electrodynamics the capacitance and the coil reluctance magnitude is **in**variable (independent) in respect to the angular speed magnitude. In comparison, the **SRT** comparable entity of inertial mass reluctance is **believed** (Einstein was compelled to believe) to emancipate by the Lorentz factor to infinity at the speed magnitude of **c,** when accelerated from a fixed point and also when accelerated mutual reciprocally in accordance with sample#5 formula#8, #8a. Additionally, **SRT** arbitrarily (without proof) assigns the infinite emancipation (the ballooning) of the inertial mass reluctance characteristic by the Lorentz factor, magnitude at the speed of light, to all backrest types of inertial mass motion constructs independent to the nature of the backrest applying to the acceleration or the time duration of the impulse; while the Hadron accelerator speed gain limitation at **c** only applies to the angular (circular) acceleration not the vertical / axial (up / down) acceleration used for the beam collision trigger aim. Why are we accepting the derivation of a feedback loop applying to inertial mass motion when the V_{origin} approaches a 10^{-9} magnitude wherein the impulse time duration over 1 meter distance diminishes to a miniscule **3.3 Nano seconds,** this **Nano second** range time duration magnitude is **only** applicable to the zero rest mass of Photons appearing from a photo diode! Most electronic scopes test equipment has a maximum resolution of 100 Nano seconds=0.1 micro seconds=10^{-6} second, why should a macro inertial mass notion in size of electrons / protons respond to such a "miniscule Quanta" in the first place? Furthermore, in congruence with macro electro dynamics: a power transformer is unable to respond to radio signals in the micro second range, its response range is in the centi 10^{-2} range. Accordingly, this Author postulates a Quantum Mechanics (granularity) related explanation for the velocity gain disappearance of an inertial mass at **c** based on the "inherent granularity of the

"quanta of light, the granularity of inertial mass and energy",

wherein p=Fdt=mdV diminishes to Zero because **dt** cannot diminish below an inertial mass Time constant:

$$dt_{minimum}=1m/c=3.3356*10^{-9} \text{ seconds duration,}$$

it is an absolute constant for an acceleration exerted from a **fixed point**; the force is not empowered to follow the acceleration beyond $V_{orgin} = $ **c**.

Let us look at the extreme electron gun in view of Galileo's equal notch section distance **s:**

$$V_{origin}=c$$

Let $V_{,new\ speed} =(V_0+V_{, gain})$; then:

Formula **#1** is: $F_{force}=m((V_0+V_{,gain})^2-V_0^2)/2s$;

then: $F2s/m=2V_0V_{,gain}+V^2_{,gain}$; therefore $V_{,gain}$ is very small at 10^{-8} range;

$$(V_0 + v_{,gain})^2 = F2s/m + V_o^2; \text{ then } (V_0 + v_{,gain}) = (F2s/m + V_o^2)^{\frac{1}{2}};$$

Then we arrive at: Formula **#1.5:** $\mathbf{v_{gain} = (F2s/m + V_o^2)^{\frac{1}{2}} - V_o}$.

If we consider: **F**=1Newton, **s**=½meter, mass=1Kg, **c**=V_0=300000000 m/s, then we arrive at a velocity gain of:

$$\mathbf{V_{,gain} = (1*2*\frac{1}{2}/1 + c^2)^{1/2} - c = \underline{ZERO} = 0.}$$

This is achieved for a 1 Newton force exerted in a straight line over ½ meter length against 1 kg mass exerted from a fixed point. Therefore, the mass emancipation of momentum d(mV) used in the SRT derivation is redundant; it is a form of double emancipation, one emancipation of mass by the Lorentz Factor and one emancipation based on Ke content from formula #1.5. The question we are trying to answer: Is it possible to use only the **SRT** emancipation of mass by the Lorentz factor for IP Machines when they exceed speeds greater than **c** without using formula #1, the answer is still unknown because the Hadron particle beam reaches tangential vector velocity of 99.9999991% of the speed of **c** in relation to the apparent fixed position of the Earth inertial frame of reference, which is **c**-5 meter/second; accordingly, the combined Earth orbit+Beam speed is **c**+30000-5 meter/second! This is because we are able to measure / calculate the Earth orbital speed and the Hadron particle speed simultaneously (at the same instant) and make a direct instant comparison to the Sun apparent fixed state and conclude / proof that the beam speed and the orbital speed are indeed a sum of combined speeds! This denial of the simultaneous existence of the Earth Orbital motion speed and the Hadron inertial mass beam speed is equal to the denial of the orbital motion of the Earth in the Geocentric world view; this denial of Earth / beam inertial velocity addition places our science back into the dark ages of the Geocentric Universe!

However, not every reader might follow the presented derivation from formula #1 into #1.5. Accordingly, a graphical vector-kinematic / proof type picture is provided on the next page to illustrate the diminishing progression of V_{gain} to zero in respect to growing V_{origin} plotted in respect to a **steady repeating** potential mass motion energy quantity of **X**.

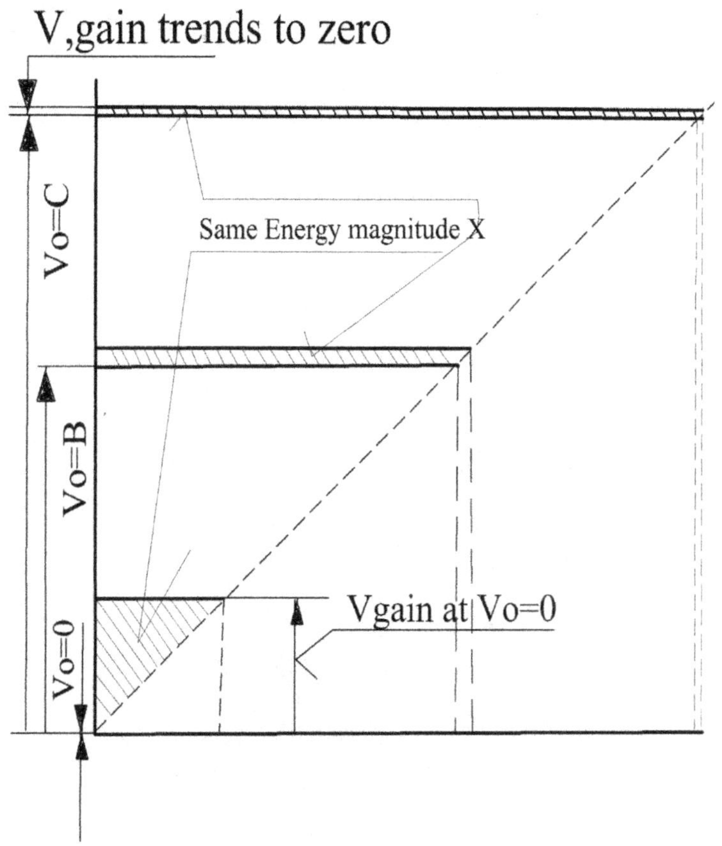

V,gain trends to zero

Same Energy magnitude X

Vo=C

Vo=B

Vo=0

Vgain at Vo=0

The V_{gain} to V_{origin} relation progression for a given quanta of energy needs a logarithmic based graph of the velocity gain magnitude in respect to the origin velocity in a range from 1m/s to a very large velocity origin magnitude of $V_o = 1*10^{10}$ m/s (10 billion meter per second) to presented the full possible range; wherein the velocity gain sinks to zero at a very large velocity origin of $1*10^6$ m/s = one million meter/second. The total energy invested into this inertial mass motion is the sum of the delta Forces passing the delta distances, the line integral of force in respect to the average time durations.

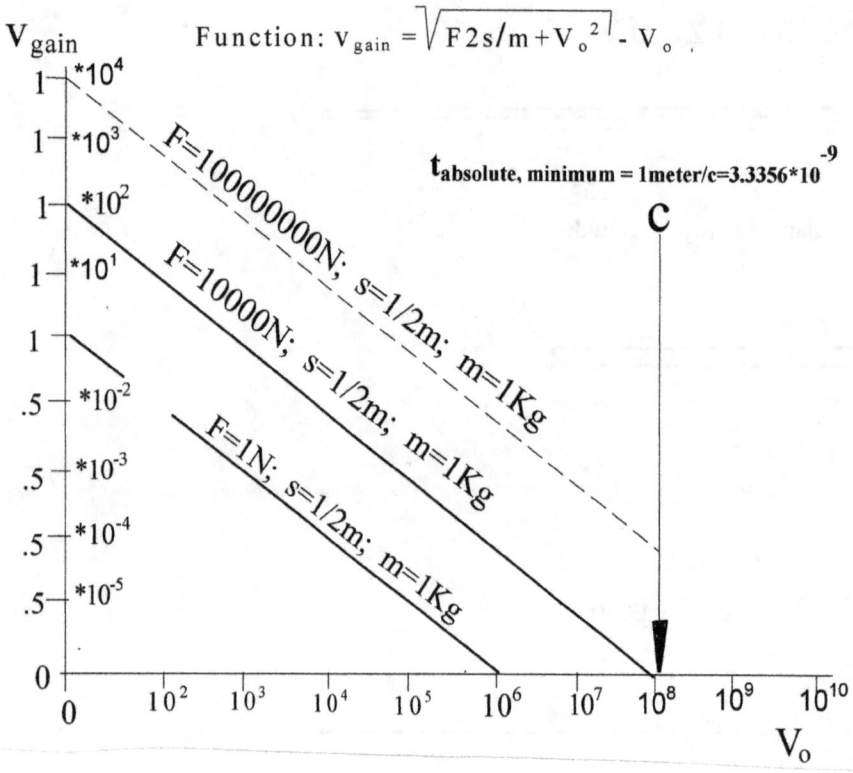

Function: $v_{gain} = \sqrt{F2s/m + V_o^2} - V_o$.

V_{gain}

$1 - *10^4$

$1 - *10^3$ $F=100000000N;\ s=1/2m;\ m=1Kg$

$1 - *10^2$ $t_{absolute,\ minimum} = 1meter/c = 3.3356*10^{-9}$

$1 - *10^1$ $F=10000N;\ s=1/2m;\ m=1Kg$ **C**

1

$.5 - *10^{-2}$

$.5 - *10^{-3}$ $F=1N;\ s=1/2m;\ m=1Kg$

$.5 - *10^{-4}$

$.5 - *10^{-5}$

0

0 10^2 10^3 10^4 10^5 10^6 10^7 10^8 10^9 10^{10}

V_o

Accordingly, at a **relativistic** velocity origin magnitude of V_o=10 billion meter per second, **much more than** an incomprehensible / unreasonable magnitude of **100 million** Newton Force pushing on 1Kg inertial mass, a distance of ½ meter in a straight line exerted from a fixed point is required to receive any velocity gain at all. There is no material stencil strength available for such a machine construct to exert such an exotic energy magnitude! Is there a projectile material available which is not being crushed into a cloud of atoms by such an immense impulse magnitude? So, **why argue about it at all in view of IP?; why are we spending billions of University dollars supported by tax revenue on an impossible construct?;** in contrast, at a **non-relativistic** velocity origin of V_o=1 million meter per second ($1*10^{-6}$ m/sec) only a 2 Newton additional force is required to receive a **tiny velocity gain;** this is then not far off from the SRT derivation; do we really need to bridge the difference factor in the range of 10^{-8} to 10^{-9} with SRT and pick up all its 9 controversies ? Accordingly, the Author is logically entitled to postulates that: **"At the speed of light the velocity gain in response to an impulse from a <u>fixed point</u> becomes zero because the inertial mass point unit is unable to respond, is unable to obtain a frequency / time response, to an impulse shorter than a <u>time constant</u> of:**

$$T_{minimum}=1meter/c=3.335640952*10^{-9}=3.336 \text{ nano second.}$$

It is experimental fact: Inertial mass point size units are immune to impulses shorter than $T_{minimum}$ **from a fixed point;** this is caused by the flexibility of the inertial mass, it temporarily deforms in response to a **hyper short shock impulse** before the **whole** mass gains momentum, this is the difference between a tennis ball and a cannon ball. The electron behaves inside a Radio vacuum tube like a tennis ball; it bounces off the anode surface when accelerated by the high anode plate voltage applied. This problem with electrons within radio tubes was not solved until 21 years past the SRT formulation and not considered in 1915. The fact that electrons are quite springy, or eve fluid, is a very good reason / explanation for the rejection of impulses shorter than $3.3*10^{-9}$ seconds. This rejection is congruent with the frequency response limit in all other Physics / Math domains and it avoids the difficulties of Einstein's electrodynamics of inertial mass motion leading to an actual incongruence with the parameter progression of electrodynamics, the difficulty of mass emancipation, the mass size contractions, time dilations and 8 unreasonable paradoxes. The Time, Space and Energy fabric of SRT and GRT accordingly, is still used with the **limitation of the impulse time duration at** $T_{minimum}$, wherein the fabric dimension is $T_{minimum}$ over one (1) meter space interval; this is a clean derivation in comparison to the founded Lorentz factor SRT derivation. However, in both, Special Relativity and in Galilean Relativity the physics within a moving inertial reference frame platform (like the Earth) remains unchanged at any speed, wherein **c** remains **c** when the moving platform reaches **c,** V_{origin}=zero and time=time of the port of departure is indicated by an electronic complying chronometer clock.

The **most precious** moving platform is, of course, the Earth with a platform speed of $3*10^{4}$ (thirty thousand) meter per second and an inertial mass reference frame V_{origin} of Zero applying to all inertial mass motions on top of the Earth surface; independent to the orientation of any inertial mass acceleration, perpendicular or in line to the $V_{orgin,Earth}$. The combined tangential velocity of the Hadron particle beam and the Earth tangential velocity is: **c-5+30000** meter / second in relation to the inertial mass centre of mass of the Solar system when the Hadron electron beam motion is in-line with Earth orbital motion, and a momentary tangential velocity of $3*10^{4}$ **-c-5** meter / second in relation to the centre of mass of the solar system when the Hadron beam is opposite the Earth orbital motion; however, **no variation of velocity** due to the Relativistic emancipation (the Lorentz factor product) of the mass reluctance magnitude to infinity has been recorded at $c-5+3*10^4$ meter/second. Therefore, it has been proven that the electrons / neutrons within the Hadron accelerator **do** (still) **respond** to an acceleration at an absolute speed of $c+3*10^4-6$ meter/second; accordingly, the emancipation of inertial mass reluctance by Lorentz factor at **c** to infinity within SRT has been proven to be redundant in relation to **IP** and the absolute speed limit of an inertial mass moving platform is larger than **c** in reference to the solar system center of mass (CM). This is because: the whole solar system is in a **highly defined fixed inertial mass orbital motion Kinetic Energy relation** and not in a **SRT** visual velocity addition relation, because the knowledge of the light speed visual aberration cancels out in view of the knowledge of the inertial mass motion relation and the General Relativity Theory spacetime curvature does not apply for the very limited dimension of an **IP** device. Accordingly, we are then able to postulate within this book: Inertial Propulsion **is** proven as valid and also, the unlimited speed limit of an Inertial Propulsion Vehicle moving platform is proven and then the platform clock Time=Port of departure clock Time has been proven, because the root cause of the **SRT** velocity gain-decline to zero applies only to the exertion from a **fixed point** not a mutual and reciprocal exertion within an inertial frame moving platform related to **IP** and applies to formula **#8, #8.a** of example **#5**. To depict this process in concert with the "Oberth Principle in page 159" a graph is provided on the next page:

Function: $V_{gain} = \sqrt{F2s/m + V_o^2} - V_o$

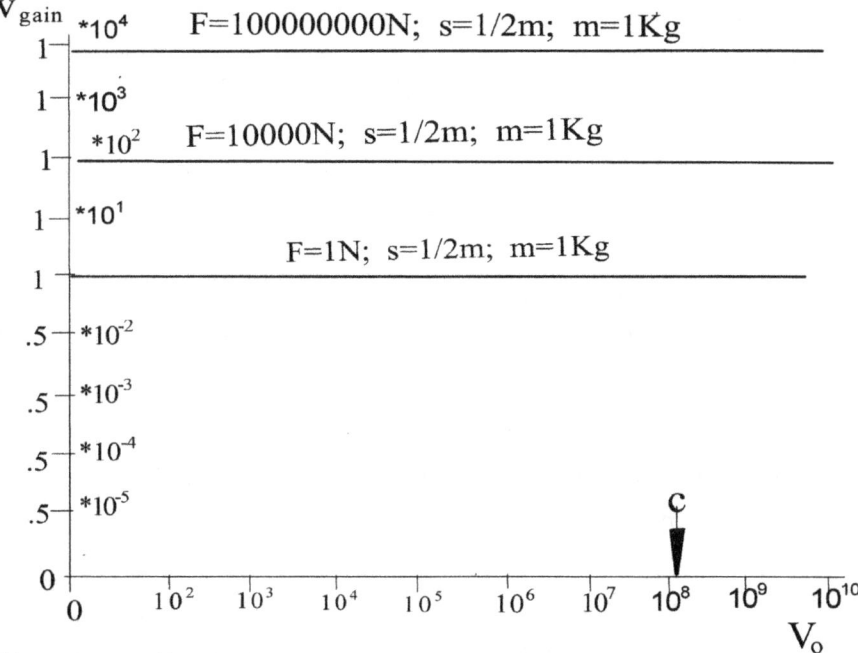

Reference: Robert DiSalle: "Space and Time: Inertial Frames",

The Stanford Encyclopedia of Philosophy

In concert of the diminishing Velocity Gain principle, the horse race is used: #1.4 Power, Hp=Force, from formula #1,*$V_{average, per work duration}$/750 Therefore, the inertial mass motion power requirement is the force multiplied by the velocity of the force.

$$\text{Power, Nm/s} = \text{mass} * V^3_{max}/(4* \text{distance}*750)$$
$$V_{max} = ((\text{Power} * 4 * \text{distance} * 750) / \text{mass})^{1/3}$$

This mean the horse is running out of horsepower to produce visible velocity gains at approximately V=60km/h over a 1km race track; the friction and other losses are again disregarded. This can be clearly noticed at the horse race. The jockey whips the horse, but according to the fence posts reference points there is very little change in the horse speed, even with the most vigorous whipping. This is what we experience in these electron gun experiments of exertions from a fixed point; the velocity gain stalls due to the MAXIMUM Ke storage capacity $Ek_{max} = mc^2$ has been reached for inertial mass motion exerted from a **fixed point** at the velocity **c**. This is an engineering-oriented, or do we want to say this is an understandable explanation of the SRT phenomenon in congruence and in formula symmetry with Huygens previously accepted original base fundamental Vis Viva mgh=E=mV^2/2 principle, without the SRT

45

paradoxes. Why is SRT not presented in this displacement domain derivation? This is because the behaviour of light; it cannot accelerate beyond the velocity of c when reflected by hyper fast rotating mirrors! However, isn't it illogical trying to accelerate light beyond c while the maximum Ke storage capacity and the internal intrinsic mass energy equivalent of an infinitesimal small inertial mass particle are also related to c? Why should a photon fly faster than c when it emerges from an atomic structure?; this question was attempted to be answered by Einstein's derivation of the electrodynamics of mass; however, we have also shown that the electrons of the Hadron accelerator exceed the relativistic mass motion velocity barrier in relation to the centre of mass of the solar system, thereby traversing the absolute maximum velocity barrier of c. Accordingly, the speed limit of an inertial propulsion system must still be proven with experiments because Inertial Propulsion works with self-contained impulses derived from a complex combination of cyclic straight line inertial mass motions translated (reflected) from rotational inertial mass motions and are **exerted mutually and reciprocally** within an isolated system **without the aid of an external fixed point**; wherein mutually means the forces are applied mutually and the mechanical first original root cause energy is reciprocally distributed by exertion between them, while the bodies having each a Velocity origin of $V_o = $ **ZERO (0)** at the start of each motion. This is a sort of accordion player motion performed by an independent suspended mechanical rotor-crank, also referred to as a complex flailing flywheel motion or a simulated inertial mass expulsion. The accordion player inertial mass motion is independent of the speed of the moving platform. It is "played" on and on with the same effort, even near c or perhaps beyond we really don't know; its origin velocity is always considered $V_o=0$ (zero) and is presented on page #73-79. This applies to Galileo's, Newton's and Einstein's invariable inertial frame of reference; the Physical Laws on a moving inertial platform are constant and independent of the speed of the platform at any speed. The speed limit of IP applies to Newton's unlimited speed limit of formula #2, if this has even the slightest uncertainty, then we need experimental proof to complete our science knowledge!

In view of Steve Hawking's intelligent design arguments, this author argument is: The intelligent designer allows billons of planets, it has encouraged building of the Noah's Ark, but decidedly disallows death rays guns made with electrons beams faster than the speed of light; accordingly, inertial propulsion beyond the speed of light is allowed by the intelligent designer for populating the trillion planets of the universe because intelligent design instructs us to **go forward** be fruitful and multiply, while our science is apparently self-limiting. In further contrast, for the time domain analysis we say that the horse is applying a measure of force multiplied by a uniform-repetitive measure of (isochronous) time duration intervals, which is the impulse-magnitude, to its legs which motivated the body of the horse to a proportional incremental higher velocity; this velocity gain is **independent** of any previous velocity **magnitudes** and independent of any speed limit.

#2B)   **Impulse, Ns = Force * time$_{delay,interval}$ = mass * Velocity$_{gain}$.**

In the time domain analysis it seems easy for the race horse to win the race, more impulse magnitude results in proportionally more speed. But obviously, the time domain analysis does not take into account how often the horse has to move its legs per each interval of the displacement. The horse is using more and more of the force effort for moving just only its legs back and forth in ever longer distance per time intervals. We can say that the time domain analysis has the disadvantage of **NOT** having a built in description of the first original cause and effect. What is causing the time based force to appear in the first place? What is causing the force to be exerted following (tracking) a hyper elevated speed and what empowers the force to follow the acceleration of the horse? Where is the potential energy, the horse's measure of oats which is causing the force to appear? Some publications also say the arguments within the time domain analysis are circular. While the time domain analysis provides the profound advantage of a proportional relationship of impulse to mass motion speed gain at any speed magnitude, it disregards the mechanical ability of the horse to deliver (the energy flow magnitude) such a mass motion impulse at an elevated speed from a store of potential energy. The time domain analysis, most importantly, disregards that the horse having the **highest** <u>**average**</u> **speed** magnitude will **win** the race, if the total race speed-gain at the finish line is identical between each horse participating in the race; this is because the time domain only analyses the change in momentum. If the horse delivers a higher force per equal time intervals at the beginning of the race, while the total sum of all the impulses remains constant, it has a higher chance to win the race. This sequence sensitivity is not possible to extrapolate from the time domain analysis with formula#2, but can be extrapolated from the displacement domain analysis with **formula#1, #1a,#1b,#1c,#1.1, #1.2, #1.3, #1.4, #1.5, #1.6,#1.7, #1.8….**

The disadvantage to correlate the impulse to the average velocity is limiting the applicability of the time domain analysis. For matter of fact, the average force per time interval delivered by the race horse cannot be calculated with impulse or momentum formula #2 until the energy magnitude is known, because the magnitude of the acceleration, the first original root cause of the motion and the root cause of the race time duration, is depending on the energy

47

expended over the race track distance:

#1.c)
$$\text{Acceleration}_{,\text{average}} = \frac{\text{Energy}_{,\text{ work, magnitude}}}{\text{Distance}_{,\text{track}} * \text{mass}_{,\text{ horse}}}$$

#1a) $E_{enrgy} = m V_{gain} V_{average}$; #1.a.2) $m V_{gain} = E_{energy}/V_{average} = P_{momentum}$

The acceleration is derived by dividing formula #1b on both sides of the =sign with the distance **S** and moving the mass down into the denominator of the energy to distance quotient. The relationship of energy and acceleration is a displacement domain/energy analysis, a uniform proportional relationship; double the energy magnitude will generate double the acceleration for the same mass and distance. The acceleration/work theorem is always true no matter how the force varies over the distance because of the before mentioned mean value theorem and the Lagrangian / Hamiltonian conservation of energy theorem, no energy quantity can be gained or lost into nothing within an isolated system. The conclusion is: The horse race can NOT be calculated or predicted in the time domain until the race is finished. The time duration is always true no matter how the force varies over the distance because of the before mentioned Mean Value Theorem and the conservation of energy theorem, no energy can be gained or lost. So, the conclusion is: The horse race can NOT be calculated or predicted in the time domain until the race is finished. The time duration is known because the race time duration itself is depending on the displacement domain analysis, an energy analysis. However, the time domain analysis within Formula #2 can be expanded by the straight line displacement on each side of the equal sign, the left and the right side of the formula, to arrive at:

#1a)   **Ke, Energy, $_{work}$, Nm = mass$_{,horse}$ * Speed$_{,gain}$ * Speed$_{,average}$**

The speed gain is (new speed −old speed);

mass * speed gain = change in momentum;

the algebraic average speed =(old speed + new speed)/ 2;

the mean value of speed= length, displacement/ time, duration

Energy work is directly proportional to the product of speed gain multiplied by the average speed of the horse wherein the mean value of formula #2 is preserved and we apparently got rid of the pesky 2 divisor for the uniform motion considerations from an initial analysis view.

Detailed, careful analysis reveals that the algebraic average speed has, again, the Cariolis **2** divisor, this means formula #1a is in reality formula #1b in disguise wherein multiple calculations are needed if the average speed is

48

derived from a non-uniform speed gain.

In contrast, if we use the mean value of speed, then we calculate: Speed$_{average}$ = distance/time. When we use the mean value of speed instead of algebraic average speed in formula #1b, then we have a diminishing certainty-accuracy level when we consider extreme non-uniform forces, because it is derived from the mean values of force and the mean value of speed.

Uniform acceleration in comparison to Non-Uniform motion

Here we arrive at the caution of the sometimes used F=ma for the integral of work performed for a variable force not changing in a straight line incline or decline in respect to time progression.

If we take a snapshot of the force

$$F=m\ dv/dt.$$

Then we multiply the snapshot with the displacement dS, we supposedly get then a snapshot of work. Within this snapshot analysis, the very small section of displacement dS is having a magnitude in relation to the magnitude of each dv and dt sections, they are the average velocity magnitude. This magnitude relation **is depending, with very great importance,** on the magnitude of speed. The displacement section is: dS=(speed, average)dt; but the speed algebraic average is:

$$V_{average}=\tfrac{1}{2}(\text{speed}_{previous}+\text{speed}_{new})$$

and **is** depending on the speed **magnitudes**.

Accordingly we can say that the three deltas dv, dt and dS sections are **not shrinkable in proportional parts** when considering high magnitudes of velocity. This is because we cannot mix the **in**dependence to the speed magnitude with the **de**pendence on speed magnitude within a term. We must keep in mind that Newton's inclined board notches have a quadratic distance interval progression in respect to the speed magnitude, the higher the speed magnitude the longer are the spaces between the notches. Then it is obvious: The integral of work must be an integral based on formula #1, #1B, #1c. It must be **in reality** a sum of all work sections applying to formula #1B. It must be a sum of Galileo's notched board displacement analysis system with very small notch distances. Alternately we can say: The sum of work is the gain in momentum multiplied by the average velocity, wherein the average velocity is defining the velocity magnitude of the snapshot time frame. This is the sum of

de=d(mv)dS/dt,in all cases it is the sum of the all the **Vis Visa / Energetic sections applying to a constant acceleration**. The difference in accuracy for Kinetic energy calculations for a variable force is presented in the Cariolis,

Formula #1b formulation and presented in contrast to **F=ma** formulation in the next picture:

Uniform versus Non-Uniform motion

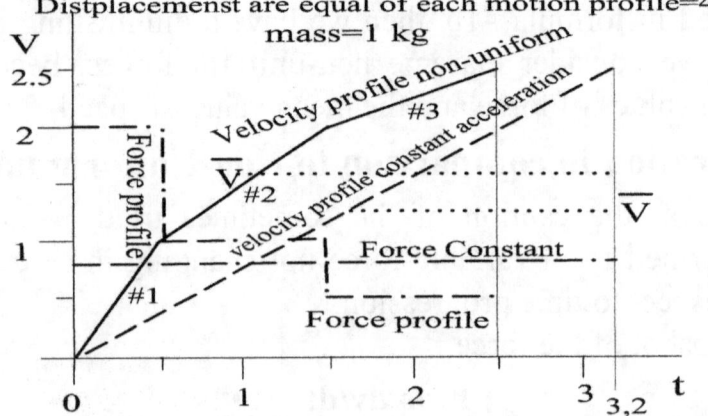

Distplacemenst are equal of each motion profile=4m
mass=1 kg

Comparing results for non-uniform motion:
Cariolis work/energy is:

$$Ke=m\tfrac{1}{2}(dV_1+dV_2+dV_3)^2 = \tfrac{1}{2}(1+1+0.5)^2=3.125 \text{ Nm}$$

$$Ke=m(dV_1 s/t+dV_2 s/t+dV_3 s/t)=0.5+1.5+1.125=3.125 \text{ Nm}$$

The mean value work/energy using average a and speed:

$$Ke, medium=(dV\#1+dV\#2+dV\#3)S/T=m(1+1+0.5)4/2.5=4 \text{ Nm}$$

$$Ke, medium=ma*S=dV/dT*S=(m2.5/2.5)4=4 \text{ Nm}$$

The mean value logic of formula#4 seems to suggest the possibility that steady cyclic repeating speed gain amplitudes and a variable average speed per race track distance, considering only a straight line displacement mass motion, could produce a directional difference in impulse magnitudes when considering continuous cycling horse races in opposing track directions. Within such a continuous cycling horse race from the starting gate to the finish-line and then back to the starting gate in an easy constant acceleration, then the higher average speed will perform more cycles and therefor more impulses and higher energy flow per total time, even though the speed gain is always repeating.

Here is a picture of this concept:

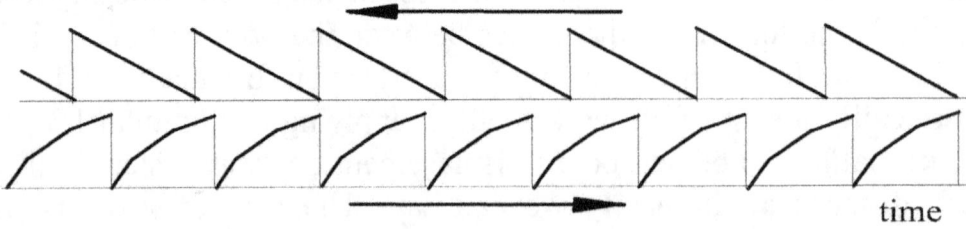

time

In the previous picture we have 9 impulses from the starting gate to the finish line and 7 ½ in the back direction to the starting gate on the left! This difference of serial impulses in respect to time is an important universal principle in physics evident in electrodynamics, Light dynamics and thermodynamics; it is the microwave cooking principle of hyper frequency impulses with steady amplitude and the principle of red / blue light impulse onto the solar-cell. This principle earned a young Albert Einstein a Nobel Price, but where is it in our physics books today? The difference in impulse will be analyzed with an opposing conveyor belt system and the vertical impulse of the pendulum, thereby proven to be correct. This correlation However, further analysis proves also, **purely straight line** displacement systems, working with divisible conveyor type mass combinations do not and cannot produce a working independent Inertial Propulsion system as correctly postulated by Newton's Third Law. This postulation will be again analyzed with reiterations of different mass motion combinations when considering mutual reciprocal straight line motions on a frictionless surface and will be found to be also true. This limitation is applying to purely straight line displacement motion of the horse race. It can also be extrapolated by analyzing the finish-line photos of a horse race. If we take consecutive photos at equal time intervals as the horses approach the finish line, it will show that the speed of all horses are in most cases identical. This means the momentum of each horse is having an identical momentum when we assume that the mass of each horse is identical. Accordingly, each race horse received an identical sum of impulses. This now seems a paradox as each horse is showing, in the photos, a different distance to the finish line. Yes, here we must again point to the difference in analytical capabilities of displacement domain analysis versus the time based analysis. If we consider the case of the most eager horse in the race attempting to accelerate a few seconds before entering the finish line and actually manages to move up in position only 10 cm short of an equal position with the lead horse, then we can say, "The eager horse has performed a higher impulse sum and has acquired a higher momentum as the lead horse but is still not winning the race". This is because the eager horse needs an advantage in acceleration to catch up with the lead horse position, then displacement multiplied by acceleration is an energy consideration applying to formula#1c.

Then we are able to postulate:

Comparative, mass motions having equal position and equal time durations can have **unequal** impulse - momentums. This is what we are trying to accomplish with IP; to accomplish IP we need unequal directional impulses within one repeating cycle.

To positively identify the limitation of the time domain analysis a plot of the horse race is presented plotting a constant acceleration compared to a variable accelerating race horse is depicted next. The distances and the separation between race horses is the geometric area under the speed contour curve above the time line; as depicted in the graph:

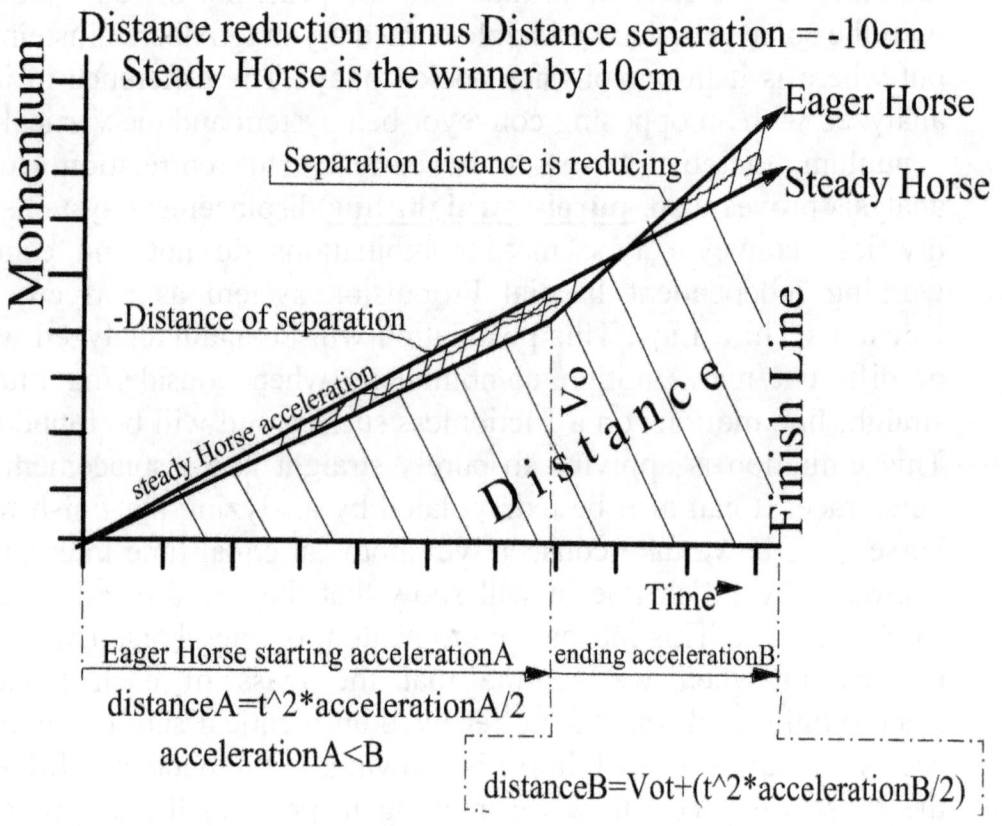

Distance reduction minus Distance separation = -10cm
Steady Horse is the winner by 10cm

The graph surface areas of distance of separation and distance of reduction can be equalise to accomplish the simultaneous arrival of the horses at the finish line; here we arriving at a possible IP requirement of reciprocal unequal impulses delivered by each horse combined with equal race time intervals. The time base analysis does not provide us with a **practical** way to answer any distance questions related to time, or allow us to formulate a practical winning strategy based on stop watch readings **before the race** has started. Only when applying involved integration of all the instant speeds, after the race is finished, can we correlate the sum of all the instant speeds to the horse position per time. This integration cannot be performed before the race because the progression of the racehorse speeds, accelerations and total race time durations are unpredictable. However, such a velocity integration is actually a displacement domain analysis in disguise, because the

$$\text{distance, }_{delta} = V_{average, per delta}t_{ime,delta};$$

wherein the sum of all the delta distances is the total distance.

Then the total distance is for constant acceleration is:

$$S_{section,distance,acceleration} = V_{average} * t_{duration} ; V_{average} = \tfrac{1}{2}at ; s = \tfrac{1}{2}at^2.$$

Accordingly, the usually presented **s = vt** pertains only to **one steady speed** or to the average speed $=(V_{origin}+V_{New})/2$; then

$$S=(V_{origin}+V_{New})t/2$$

applies to **constant accelerations**. Furthermore, the integral of impulses cannot provide us, in any way with an answer to race horse position at time duration; it only provides us with a momentum magnitude. In contrast, the displacement domain analysis of formula #1, #1.1, #1.2 #1.3 provides us with a position analysis of each race horse with the possibility to extrapolate to a minimal race time duration by co-relating the potential energy, to work magnitudes, to the average speed per race track distance markers. The speedometer technology, in every person's car, is based on the **distance markers** in relation to the straight line displacements coupled to the angular motions centrifugal forces strength allowing the motorist to gage his acceleration magnitude. This, accordingly, presents the highest efficiency of thought for isolated system machine inertial mass motions. This is how Christiaan Huygens solved his pendulum problems between the years 1659-1673, up to 14 years prior to the publication of Newton's first Principia. Then we must ask: "Did Huygens know about the impulse to momentum limitation, and importantly, did he **need** to know the impulse to momentum relationship to solve his oscillation problems?" Yes, Huygens knew about the impulse to momentum correlation which he helped to formulate with Descartes. No, Huygens choose #1, #1.1, #1.2, #1.3, #1.4, #1B and was successful in finding the pendulum time interval formula: $T=2\pi(L/g)^{\frac{1}{2}}$. His derivation process using the Lagrangian **L=T-V**, herein the maximum mechanical potential of the pendulum, the maximum potential energy V is the height in form the pendulum length L, the potential energy P_e minus the kinetic energy K_e is the constant Lagrangian and the radius of orbit is equal to the pendulum length:

r=L; V_{origin} =0; acceleration is g=9.8m/s:

Formula **#1.d**: $V^2_{gain}= 2gL$; **Vis Viva$_{,average}$=gL**

Convert to angular form: $\omega L=V$; $\omega^2L^2=V^2$

Insert the angular form into formula for Vis Viva:

$$L^2\omega^2_{average}=gL \ ; \omega^2_{average}=g/L$$

Breakout the Time=**T** from the angular form: $\omega=2\pi/T$; $T=2\pi/\omega$;

Insert **T** into the angular form of Vis Viva;

$$T=2\pi(L/g)^{\frac{1}{2}}$$

This is called **Huygens Pendulum formula** from **1673**.

Newton worked on the pendulum problem in his second Principa Book edition, published **40** years past Huygens pendulum formula, in anno **1713**

using a difficult verbal style force, impulse, mass, velocity and distance relation; an analysis starting in the time domain and progressing into the displacement domain: The Proposition **XXIV XIV** in a copy of his own Principia writing:

For the velocity which a given force can generate in a given matter in a given time is as the force and the time directly, and the matter inversely. The greater the force or the time is, or the less the matter, the greater the velocity will be generated. This is manifest from the second Law of Motion. Now if pendulums are of the same length, the motive **forces** in places equal **distance** from the **perpendicular** are as the weights: and therefore if two bodies by oscillation describe equal arcs, and those **arcs** are divided in **equal parts**; since the times in which the bodies describe each of the corresponding parts of the arcs are as the times of the whole oscillation, the velocities in the corresponding parts of the oscillations will be to each other as the motive forces and the whole times of the oscillation directly, and the quantity of matter reciprocally: and therefore the quantities of matter are as the forces and the times of the oscillation directly and the velocities reciprocally. But the velocities reciprocally are as the times, and therefore the times directly and the velocities reciprocally are as the squares of the times; and therefore the quantity of matter are as the motive forces and the square of times, that is, as the weights and the square of times. Q.E.D.

Apparently, Newton effectively switches half way through his derivation from time domain analysis to displacement domain analysis by using the square of time in union with equal spaced notches of Galileo's notched board; he uses equal spaced displacements at this later year of 1713. This is because time duration is $t=s/V_{average}$ includes the displacement **s**. Within the time domain analysis the time is $t=V_{gain}m/F$ derivation from impulse$=Ft=mV_{gain}$, it has no ingredient of the displacement**s**. Here he effectively invents the differential equation for oscillations and shows the congruence of his formula with Huygens / Leibniz Vis Viva Energetic principle. While he delegated the angular reflection (translation) from straight line impulses out of his **first** Principia Book edition, as presented in page 20. The anno 1713 time domain derivation of Newton's pendulum requires a difficult final calculus integration step simulating the angular quantity ω for arriving at the coveted pendulum Cycle Time interval formula; this step is usually omitted and Huygens formula inserted as the final solution. However, Newton's derivation method allows us to use progressive smaller arc motion sections of the total pendulum motion arc, called the displacement analysis in respect to time $f(a)=d^2s/dt^2=\omega^2s$ in calculus terms, arriving at an **logical proof analysis** in respect to impulse / momentum magnitude; therefore, this publication is compelled to use Newton's method of the horizontal **distance** to the **perpendicular** in terms of the pendulum length giving the sine ratio, accordingly arriving at the final absolute prove of the validity for continuous cycling inertial propulsion. However, Newton did not prove with his Pendulum derivation how it applies to his original exclusion of these kinds of motions from his third law presented in page 20; does the third law apply or not apply, in every situation. We will see with centripetal energy transactions mixed with straight line impulse transactions mixed with complex energy transactions, the third law in the impulse – momentum form does **not** apply in every situation, but, the third law in the energy form applies for IP type mechanisms! From these points of initial analysis we can postulate already with certainty: Machinery like the Pendulum, Race Horse, the Indy 500 Car racer, Planetary motions, the Inertial Propulsion or any other machinery, where position in relation to time progression is important, must be analyzed **first** in the displacement domain. This is, because, it is not **practically** (simply) possible to exactly extrapolate the sum of impulses and the resulting momentum to the initial potential mechanical energy first original root cause of the motion. Machinery internal inertial mass motion is an energy management problem, while the impulses are the proportional reflections of the momentums. Here we arrive at the **profoundly important** physics postulation for machines:

The first original root cause of inertial mass motion within machines is the exertion of a work quantity taken from a quantity of potential energy at each displacement position, causing a gain in speed for each change in position, causing an accumulative average speed at each position in relation to the initial starting position, causing an accumulative total momentum and causing the total motion time duration over the total accumulated displacement distance. Accordingly we can conclude: Within machines, the directional energy flow is the originating fundamental working principle.

This is James C. Maxwell's postulation:
"All phenomena depend on the variation of energy";
and now, we able to add:
"including the phenomena of Inertial Propulsion".

For someone stubbornly maintain that all inertial mass motion problems are solvable with formula #2, without any actual real displacement length parameter considerations, we are certainly entitled to comment: You surly are disregarding the practical necessity for the five reformulation of Newtonian mechanics from singular time domain analysis **back** into the displacement domain analysis; you are ignoring **the universal reach of energy conservation**

within machines and the commercial success of the refinements applied to Huygens / Leibniz "Vis Viva" principle leading to the, Lagrangian, Euler, Hamiltonian, Sir Kelvin's, Coriolis and Hertz inertial mass motion energy principles having its original root in formula #1.

The common notion that any proof of the limited reach of the Third Law must be automatically deemed invalid is accordingly traversed. This principle will be proven to final exhaustion with many examples. To complete the range of analysis of IP must include all possible changes in variables and we must also include the analysis in the frequency domain, the play of forces in relation to a change in cycle frequency, the **FM** modulation. This is because the presented IP system works with the variations in cycle frequency.

Reference: **www.physics.int/motion-graphs/**

The Invention of Inertial Mass Kinetic Energy

The scientific **re-**investigation of kinetic energy was urged along by the emergence of the steam power technologies. Engineers were compelled to provide a simple answer to an economic question: "How much fuel does it take for a steam engine to propel itself from point A to point B at an **average speed** $V_{average}$?", wherein the highest possible average speed to the least magnitude of fuel consumed is the most desired condition; here again, we do not consider friction losses. It was easily observed that there was a content of fuel quantity contained within this machine inertial mass motion **escalating** with the magnitude of speed $V_{average}$, a fuel quantity unrecoverable because of the discarded heat during breaking. It was also observed that Newton's force was present in form of the boiler pressure acting onto the piston area providing the force for the machine motion acceleration. It was easily observed that the acceleration was disappearing at elevated speeds with a constant boiler pressure applied and thereby assumed constant motive force applied! It was **rediscovered** that the correlation of the fuel quantity question and the top speed / average speed of these machines were all caused by the Physics of formula#1 and its direct proportional relation to the energy based gas (steam) law. The steam cylinder pressure is inversely proportional to the steam cylinder volume $P_0/P=V/V_0$; more cylinder volume caused by the longitudinal piston motion reduces the pressure for the same amount of steam. The caloric law provides the proportional relation of gas energy to the cylinder pressure; more calories will produce proportionally more steam pressure and more acceleration based on formula #1. Accordingly, formula **#1** provides us with the fundamental **symmetric Physics foundation** of machine / mechanical inertial mass motion.

Within the steam driven machines the flywheel was used as a motion dampening device, in an observable congruent function to the fuel question. In a simpler, but also related case, the question arises: "Is it possible to solve a compressed spring expansion problem or any other problem involving stored potential energy without using the work, which is the force times distance expended loading the spring?". Is it possible to use impulse, momentum and equal reaction to an action or acceleration analysing the play of forces in the time domain alone? NO SORRY, impulse can only be stored as mass motion momentum because the passage of time quantity disappears into history. Only work/energy, the force times the distance can be truly accumulated, mechanically stored, compared to other forms of energy and reconstituted into impulse which will be proven next with certainty using the Swiss Gyrobus example. Therefore, the statement of "EVERY" or, "is it ALWAYS?" in Sir Newton's third law must be carefully examined. In most good physics books the third law is restricted to objects of very small dimensions (point mass) motivated by the exact coincident of single shot-put straight line vector forces. The net RECOIL of a complex mechanism working with the principle of kinetic energy transactions, like the Gyrobus or the presented IP drive, **cannot** be solved or calculated or described with Newton's impulse, momentum, reactions and accelerations in the time domain **alone**. IP requires FOREMOST work and kinetic energy of the two vector dimension of mass motion in the displacement domain; this is because the centrifugal force and the inertial mass motion are both in the displacement domain, they are congruent entities; they both have a displacement entity in the denominator. This is the scientific work of Galileo, Huygens, Leibniz, Lagrange, Euler, Hamilton, Cariolis and Lord Kelvin. G. Carioles made accurate measurements of the relationship of mass motion kinetic energy in relation to the friction heat an inertial mass motion break action delivered. This is how he found that the friction heat energy magnitude has the equivalent of $Ke = \frac{1}{2}$ mass * Velocity^2, gain. The $\frac{1}{2}$ is trying to tell us that inertial mass motion progresses in uniform time based increments in correlation to the gain in friction heat energy, wherein the impulse calculation will be shown with formula #6 on page 68. This seems to reinforce that straight line displacement mass motion cannot perform Inertial Propulsion only by itself. "Is this how Newton proved his third law?". Fundamentally: The prime motivating first original root cause of a mass motion action is the work performed on the mass, which causes the subordinate intermediate action of impulse which obtains the end result, which is the kinetic energy of the mass. Without work performed there can be no impulse and no end result of kinetic energy. From the presented facts we can postulate the assumption of Newtonian Physics: The Newtonian relationship of uniform gain in mass acceleration in respect to a uniform force applied assumes that the energy source itself causing the mass acceleration is not being depleted or its source energy flow diminished by the energy flow demand of the mass acceleration. Furthermore, that the energy flow is not causing a negative feedback loop, therefore, not causing motions in the complex Cartesian matrix geometric plane. Then we must ask: "Does Newtonian Physics violate the conservation of the kinetic energy principle or the cause and effect principle, the negative feedback principle or the complex plane principle?". Newtonian mechanics apparently ignores every one of these sciences; because at Newton's time complex analysis was unknown. The todays Engineer has to sort out, how a system transfers energy when mass motion acceleration takes place and whether or not it generates self-contained motivating impulses in view of the complex plane. Newton's equal momentum reaction to an impulse action does not consider kinetic energy magnitude transfer, depletion-flow, cause and effect, mutual separating and combining of masses, negative feedback and complex plane implications. Subsequently, the third law ought not to be cited as contrary to inertial propulsion. Newton himself stood clear of extending his third law to rotational motion translated onto straight line motions; later we will present six exceptions of the third law Newtons' restriction is based on. Physics is after all, foremost, the study of energy and matter. A voice of reservation to Newton's time based Impulse/momentum physics came from the inventor of the pendulum clock, Christian Huygens. Huygens already suggested in 1668 AD, 19 years prior to the publication of Newton's laws, the importance of the product of force applied over a distance in respect to inertial mass motions. Christiaan Huygens' presentation was made during a London Royal Physics Society meeting clarifying the principles of inertial mass motions?. A further strong voice from the force and displacement camp was Robert Hooke who postulated Hooke's elasticity law, constant k=force/distance. Hooke's law is indispensable when analyzing systems having cyclic oscillations. Is it possible to engineer Christiaan Huygens's pendulum clock with impulse / momentum alone disregarding the periodic potential energy of the pendulum weight * height? No, height is included in the form of the pendulum length, because, the clock engineering needs the potential energy of the pendulum weight height as the root cause of the mass motion, as we have demonstrated in all previous examples. Still today, many Physics textbooks separate the time domain analysis of impulse/momentum and the displacement domain analysis of work/energy like fire and water, while both time and displacement domain analysis have an isomorphic symmetry. This means, we apparently do not have the best workable tools available for Physics of mass motion if we are not instructed to directly and easily correlate the two analysis systems. Importantly, within machines potential energy, kinetic energy and momentum coexist side by side, they are inseparable joint by isomorphic symmetry! In contrast:

The most prolific,

bestselling, best liked, solid **5 star** review and massive worldwide language edition Physics-Engineering formula book, "Kurt Gieck Engineering Formulas", has a combined Kinematics section. The Kinematic Section places the time domain and displacement domain analysis side by side for easy comparison, wherein formula **#2** is on the left and formula **#1** is on **the right**, connected with an equal sign. The velocity parameter **V** and V_0 are the **same** quantities on the **righ**t and on the **left** side and they return the identical acceleration magnitude in a **vector-kinematic** analysis.

$$\mathbf{a} = \text{formula \#2} = (V - V_0)/t = \text{formula \#1} = a_z = (V^2 - V_0^2)/2s.$$

Wherein **t** is again: $t = s/V_{average}$, $V_{average} = (V + V_0)/2$; and $V_{gain} = V - V_0$
And here in angular form:

Formula#2.6) $\alpha = (\omega - \omega_0)/t$ = formula #1.6 = $\alpha_z = (\omega^2 - \omega_0^2)/2\varphi$

Point **#1**: **Importantly:** This identifies the derivation of the average velocity, one halve of the $V + V_0$ **magnitude sum**, as the **bridge** between the displacement domain analysis into the time domain analysis and it assigns the right formula as a complex **z** domain formula. The **bridge** effectively expands time into two parts: Distance and ½velocity.

Point **#2**: **Importantly**: The choice between the right and left formula depends on the readily available "**s**" parameter; the derivation of the pendulum for Huygens and Newton was the displacement domain analysis on the right, because the pendulum bob height or the horizontal distance was used for the derivation. The present IP system has its root cause also in the pendulum and it employs complex functions, accordingly the right formula must be used and is indeed used in calculus form in many publications. The correlation of displacement analysis to time domain analysis based on the average velocity bridge appears to be the modern current state of World Physics in congruence with H. Hertz admonition that we need to correlate impulse to kinetic energy. We must not assign unprovable superior to inferior priority assignment to one particular formula over another formula without complete exhaustive analysis proof of the repercussions and in view of new technology inventions. We must constantly compare the original root cause with the subsequent action result, the degree of motion freedom / displacement; also look for time delay, motions in the complex domain between the original root cause and the action result. We must always keep Galileo's and Newton's notched boards' side by side, keeping an equally importance relationship in mind. The <u>original</u> root cause of inertial mass motion is always energy flow, because it allows us to view all issues together, complex, noncomplex and temporary time delayed; or has anyone identified a more complete, an all-inclusive efficient analysis sequence?

The analysis of dynamic mass motion using the work-kinetic energy principle has a long history, from Galileo, to Huygens, Leibniz, Lagrange, Euler, Kelvin, Hamilton, Hertz, etc. Hertz presented through his publication "The principle of mechanics presented in a new form" the direction this publication is taking; that the mass motion motivating force passing a distance is proportional to the change in kinetic energy of the mass. Impulse is an intermediate parameter derived from the original root cause of the motion, which is energy. Albert Einstein attributed some of his thought processes partially on Heinrich Hertz's publication, thereby leading to the invention of relativity which has its logical origin in the kinetic energy flow of the pendulum and the energy flow of the rotational to straight line coupled motion. Let us now look at the crossbow weapon still in use by game hunters in 1668, it develops the velocity of the arrow according to Sir Newton's impulse. However, how could Sir Newton calculate the velocity of the arrow using impulse or acceleration; the product of average force multiplied by the time of applying the force to the mass of the arrow? The force drive time duration and velocity gain are the TWO unknown parameters we are seeking! The time factor is part of the change in momentum. As we have seen from Newton's pendulum derivation he would have to converted the velocity average for $V_o=0$; $\mathbf{V_{average}=s/t}$, into the velocity gain $V_{gain}=2\mathbf{s}/\mathbf{t}$. Then using $\mathbf{Ft=mV_{gain}}$ arriving at $\mathbf{t=(m2s/F)^{1/2}}$; effectively converting to the displacement domain analysis. This shows again that the time factor is more logically solvable with the acceleration to energy relation of formula #1c in conjunction with the energy formula #1a, it is more logically congruent with the universal **Force*Distance** principle, $V_o=0$:

Formula #1.7 $\mathbf{t_{,time}=(2s_{,\,distance}/a)^{1/2}}$; $\mathbf{a=F/m}$; $\mathbf{t_{,time}=(2s_{,\,distance}*m/F)^{1/2}}$

The force and displacement of the crossbow string is the potential energy of the system and are both known quantities. Both force and displacement are vector quantities because displacement has also a three dimensional direction! The three dimensional direction of the crossbow string displacement launches the arrow with one single vector dimension of impulse! A vector multiplied by a vector ought to be a vector quantity. "Why is work done onto the arrow NOT considered a vector quantity?" The process of aiming the crossbow are steps taken aligning the force and displacement of the crossbow string in a three dimensional direction with the target; denying the work done by the crossbow string the status of vector seems strange-impractical indeed and represents one of the barriers to the complete understanding of inertial mass motion applying to machine vehicular motion. The author was unable to obtain a better explanation for the necessity for separating impulse from work done and the need for separating the relationship of the impulse vector from the work done (possible vector) that is: Force * displacement distance is an area-geometric figure of two scalar vector quantities. This publication cautiously postulates that there is a fundamental process present within the operation of the Swiss Gyro-bus momentum transmission in small quantities where the flywheel momentum vector is transformed into a scalar kinetic energy value of the bus. The suggestion by Chistiaan Huygens to use work/energy, the product of force times displacement, would have required an additional step of physics thinking (or is it assumption?), to solve all the previously presented problems. Namely that: The application of a steady uniform force generates a steady uniform gain in inertial mass motion velocity, which is Newton's acceleration, because then, the displacement is ½ of the square of the velocity divided by the acceleration, which is more a geometric rising slope triangle area problem than mass motion physics problem. Therefore, we can extrapolate that for $V_0=0$; formula#1:

Force * distance = mass * ½Velocity$^2_{gain}$,

which solves the velocity parameter and sidesteps the time and the acceleration parameter ingenuously, which was discovered 100 year later by Carioles. However, the application of uniform gain in inertial mass motion velocity in respect to time domain raises the second important logical barrier to inertial propulsion, the work / energy of non-uniform acceleration versus uniform (constant) acceleration presented in formulas #1a, #1b, which is dealt with again later in this publication. Newton publicly resented the privileges of inventors riding on the coattails of his important scientific discovery, the centripetal acceleration. But then again, why blame Newton for the skill and foresight to avoid controversy and keep his laws simply in the time domain analysis and not to include work/energy nor include rotational pendulum motion combined with straight line reflections in his first Principia. When analysing motion involving only friction and change in potential energy magnitude independent of the inertial mass, then we find a further consequence of the impulse to kinetic energy relationship. We find the average force versus kinetic energy flow relationship, where kinetic energy flow is horse power or kilo-watt, which is a proportional relationship in respect to the time interval. If we double the time interval per kinetic energy flow (Hp) we will double the force available to push the car. Yes, therefore, heavy loaded trucks will increase the time interval going up the hill; they will go slow up the hill.

FORCE, average. Kgf = HORSE, power * time / distance.

Therefore: by multiplying both sides of the formula with time we get:

#2a)   **Impulse, Kgfs = Energy, gain / Velocity, average**

If the velocity gain of a repetitive **non** inertial motion is constant, then the magnitudes of the impulses are in uniform step with the energy quantity applied. More energetic energy pulses result in larger impulses, which is good news for inertial propulsion design.

Seven Machines with internal Energy flow origins and transient momentums

Here we present seven mechanical constructs examples where transitional magnitudes of momentum appears and disappears under the controlled influence of directional energy flow; this is reinforcing the principle of directional energy flow causing a net directional impulse magnitude outcome within machines. This is traversing the universal reach of momentum conservation leading to the conservation of momentum into the Energy Form and into the physics of Inertial Propulsion.

Example **#1**. The conveyor example, the flow of quantities of kinetic energy for different masses being accelerated and transported in one single vector dimension by a horizontal level conveyor belt disregarding friction losses follows: The kinetic energy expended for each item using formula #1a is:

$$Ke, Nm = mass_{,item} * Velocity_{,gain} * Velocity_{,average, per, acceleration}$$

$$Velocity_{,gain} = Speed_{belt}; \quad Velocity_{,average, per, acceleration} = Speed_{,belt} */2$$

The formula for the Ke of each item is:

$$Ke_{, item, Nm} = mass_{, item} * Velocity^2_{conveyor, belt}/2,$$

Since the frequency of items dropping onto the belt is:

$$Cycle\text{-}frequency = Velocity_{, belt}/ (length_{, item} + Length_{, acceleration})$$

Since:

$$Length_{, acceleration} = (Velocity_{, belt}/2) * time_{, acceleration}$$

The power flow for each item is:

$$Power_{, flow, average}, Kw = Ke_{, item} * Cycle_{, frequency}$$

Since $Force = V_{, belt}/ time_{, acceleration}$; we also able to say:

#1.6)   **Power, flow, average, Kw, Hp** $= m * Force_{, average} * Velocity_{, average}.$

Formula #1.5 describes a universal principle in Physics applying to any reluctance delay process and also presents the actual monetary cost to drive inertial mass motions because Kw = electricity used per time = fuel cost per time = money per time = velocity, gain per time multiplied by velocity average is a quadratic function. It takes more money to operate a conveyor in respect to higher speed magnitudes proportionally to V^2. In contrast the inertial mass motion momentum is an isomorphic (diminishing returns) relationship to money. If we find it difficult to obtain the conveyor acceleration time duration with a Stop-Watch, then it is advisable to make a Felt-pen mark at the spot where the transported item drops onto the conveyor then measure the acceleration distance with a measuring tape just before it leaves the conveyor. Now we can rearrange formula #1.6 for the displacement domain analysis applying to Formula #1 by assuming a constant acceleration we arrive at:

$$\text{Time}_{,\text{acceleration}} = 2 * \text{distance}_{,\text{acceleration}} / \text{Velocity}_{,\text{conveyor,belt}} \text{ ;}$$

$$\text{Velocity}_{,\text{average per acceleration time}} = \tfrac{1}{2} \text{Velocity}_{,\text{ conveyor}};$$
then the power flow for each item is:

#1.6) $\quad$ **Power, Kw = m $*$ Velocity$^3_{,\text{conveyor}}$ / 4distance$_{,\text{acceleration}}$;**

The shorter the acceleration distance the larger the power magnitude;
the thrust pushing the conveyor body in respectto Formula #1.5and the size=x
of the item transported is:

#1.8) $\quad$ **Thrust = Power$*$2$*$X) / (Velocity$_{,\text{coneyor}}$. $*$distance$_{,\text{acceleration}}$);**
The reciprocal impulse per belt cycle is:

#1.9 $\quad$ **Impulse = Thrust$*$2Length$_{,\text{belt}}$ / Velocity$_{,\text{conveyor}}$**

The energy flow in respect to the work / energy formula#4, the work
performed by each item multiplied by the item cycle frequency is:
Energy, flow, Kw = (formula#4) work, per item $*$ **items # per time**
This is by far the easiest way of obtaining the energy flow of the conveyor, because here we work with the mean value of speed. The kinetic energy flow of the conveyor starts at the drive motor and the kinetic energy is released when each moving quantity of mass leaves the conveyor belt. The kinetic energy quantity is reflected by the conveyor velocity. The "acceleration" part of the formula depends on the time it takes for the items dropped onto the belt to reach the same velocity as the belt. The acceleration, which is a function of the slippage on the belt and the ability of the drive motor to maintain a constant belt speed, dictates how many items can be placed on the belt one by one in a tight spacing, therefore, the total mass being transported per time interval. The frequency of items transported, the quantity of items transported per time domain, is then a function of the acceleration, which is the principle employed by the presented inertial drive. Furthermore, a decrease in acceleration time increases the quantity of continuous force Thrust exerted by the belt against each item. The interdependency of cycle frequency, energy flow and Thrust is

therefore the same for all physics cyclic flow phenomena where amplitude of the flow is constant but the cycle frequency is variable.

For example: Let us drop a new item onto the conveyor belt one by one and compare a sticky belt having an acceleration time of 0.3 seconds and a slippery belt having an acceleration time of 0.6 second, then the Thrust differential, frequency and recoil between the sticky and the slippery belt is double as large. Thereby, kinetic energy flow must be regarded as having a direction, having a source and a sink, where the kinetic energy source is the drive motor and the energy sink is the velocity of the mass of each item transported per time interval. According to the previous analysis we postulate: The kinetic energy flow is therefore identical to the flow characteristics of all other flow phenomena in physics, as in thermodynamics, aerodynamics, electrodynamics, radiation dynamics etc. and cannot be isolated as having separate fundamental physics laws. This is the fundamental principle in Heinrich Hertz's book "Mechanics presented in a new Form" This means the devices found in electrodynamics generating great avalanches of energy must be available also in inertial mass motion, in particular in combined rotational and straight line displacement.

For an electro dynamic congruence example: If we repeatedly charge and dis-charge an electrical capacitor to a set magnitude of voltage in 0.3 seconds instead of 0.6 seconds then the energy flow, in Watt will be twice as large. These symmetric relationships were explored by Heinrich Hertz in his book, "Mechanics presented in a new Form", which proves that even complex Cartesian grid numbers, irrational numbers, must exist in rotational mass motions! However, obviously, the operation of the straight line conveyor cannot yet be regarded as a suitable candidate to implement inertial propulsion. This is because of the directional congregation of items. If two conveyors having gradient belt accelerations operate in tandem opposite directions, then items will pile up at the end of the faster accelerating conveyor. This negative aspect of the straight line conveyor is then Newton's equal reaction to an action because each acceleration time frame also contains the equal reactive collision impulses of the congregated items. The question is: "Is the straight line conveyor congregation of items a universal principle in Physics or is the coupling (the translation) of variable rotational motion with straight line motion a mechanical arrangement sidestepping Newton's reaction law?". The mechanical clocks on ships suggest there is. The work/kinetic energy flow is a combination of displacement and time domain analysis because we analyze the magnitude of energy flow per passage of time. Work/Kinetic energy flow further generates the magnitude of the recoil Thrust. The operation of the conveyor clearly demonstrates the existence of the relationship of the scalar energy flow magnitude to the Thrust magnitude applied to a mass and the machine generated vector direction of the generated impulse applied to one vector dimension of mass motion. The energy flow is in an isomorphic symmetry to the impulse. Work/Kinetic energy flow analysis, thereby, side steps the unnecessary redundant analysis complexity of work performed by the motor and the impulse applied to the mass and simply converts electrical energy flow into mass motion energy flow. We send −+Kilo-watt into an isolated system and get a gain in +−Kg, Newton, meter or +−Joules or +−Kilo, calories out. Any valid IP system formula must therefore be based on the energy flow principle. In view of the conveyor belt operational formula this publication postulates with certainty: The continuing repetitive cyclic acceleration of items dropped onto the conveyor belt is generating a continuous average energy flow and a continuous average recoil magnitude of the mechanism depending on BOTH, the magnitude of the conveyor belt velocity AND, OR, EITHER the acceleration time duration of each item transported. The steady average recoil magnitude is the consequence of the continually concatenating acceleration timing pulse durations. The timing pulse durations are design criteria and is the cause producing the magnitude of the work/energy flow magnitude.

Example #2: The perfect conveyor without slippage or drag is depicted; consisting of a motor, a flywheel mounted onto the motor shaft and a crank pin with connecting rod attached to a sliding pusher pan. The inertial mass of items drops onto the pusher pan, when the crank pin and the connecting rod are in the dead center position, allowing the inertial mass motion to accelerate relatively gently:

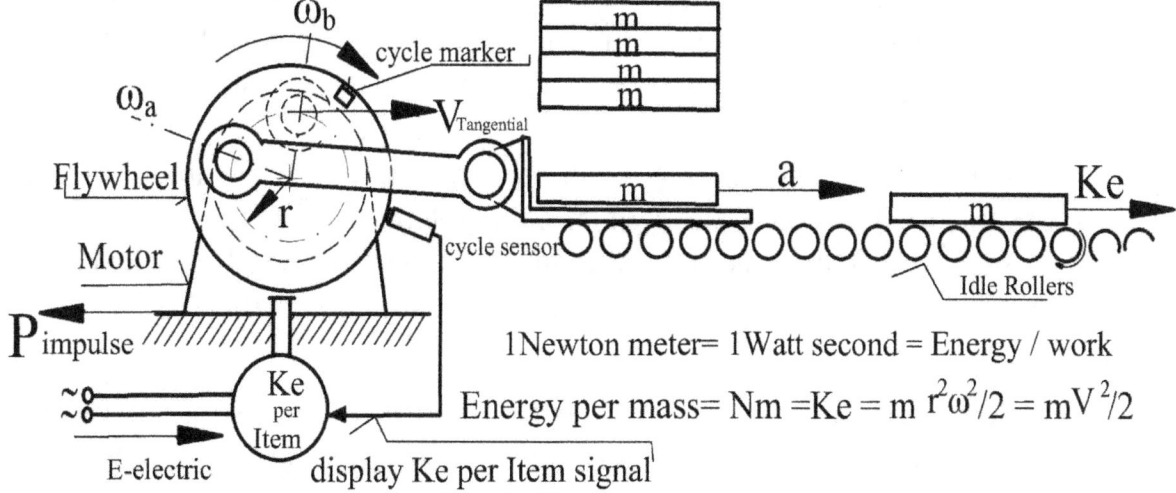

1 Newton meter= 1Watt second = Energy / work

Energy per mass= Nm =Ke = m $r^2\omega^2$/2 = mV^2/2

The Watt meter is clocked by the cycle sensor; this is updating the wattage multiplied by the implied cycle time giving steady readout of the Ke in

Nm (Newton meters) per item; wherein the Watt meter is, as usually, an averaging device giving the average energy flow in respect to the difference in angular speed $\frac{1}{2}(\omega_a-\omega_b)$ per mass transported; wherein $\omega_a>\omega_b$. This proves that the reactive impulse P is a function of the difference in angular speed $\frac{1}{2}(\omega_a-\omega_b)$, because the push against the inertial mass **m** is consuming Ke from the flywheel; this presents the primary derivation of the IP self-contained impulse. The time duration of the energy flow for one item transported is a function of the rotational cycle, one displacement revolution of the flywheel updates the watt meter and coverts the readout to Ke. This arrangement traverses the nonsensical notion that there are no Ke meters in mechanics; Ke meters are available with technology, mathematical physics has to recognise this. The display can be changed to impulse against each item by clocking the average electrical current multiplied by the cycle time duration giving the Ampere seconds. The example #3 proves six important points for inertial propulsion.

Point #**1**: Energy is the **true** first original root cause of the inertial mass motion impulse and it also allows us to solve the motions in the complex Cartesian plane.

Point #**2**: The restraints, the linear guides, of the mechanical system convert the scalar energy flow into a vector impulse; this proves the exchangeability of a scalar value into a vector magnitude by two symmetrical and simultaneous mirror image mechanical restraints systems independent of an internal rest torque; this means we need two symmetrical mirror image for IP systems; this is the Eric Lathwait's postulation of the IP requirement.

Point #**3**: The consequence of point #1, #2 is that we must convert energy flow into impulse for inertial propulsion machines by using formula #6 and #7 in page 73, 74 to arrive at the propulsion thrust; this is Newton's exclusion from his third law.

Point #**4**: The system has an upper tangential vector velocity limit based on formula 1.6, the stencil strength, and the potential energy of the mechanical
and electrical construction material, wherein the theoretical absolute extreme velocity ceiling for even the most exotic constructs using zero rest inertial mass of photons is **c**.

Point #**5**: The angular speed is non-uniform $\frac{1}{2}(\omega_a-\omega_b)$; it varies with the energy flow demand and energy supply restrains causing the IP motion in the complex Cartesian grid.

Point #**6**: **Most importantly:** The time delay, from the time the inertial mass **m** leaves the pusher tray and causes a reactive impulse against a subsequent constraint is the opportunity to remove the reactive impulse with the mutual and reciprocal energy absorbing action of example #5, formula #8, #8a; this energy absorbing action reduces the momentum into a transient quantity.

Point **#7:** The physics principle of the pusher tray motion is reversible; the same mechanism can recapture the inertial mass motion kinetic energy in reverse motion and reversed energy flow direction and return the energy into to the first original energy source using dynamic breaking! We can say the transitional momentum amplitude is **always** conserved **in energy form**!

Point **#8:** If we consider the voltage potential of the first original electrical DC energy source to be very robust and constant (stable) then the item straight line velocity amplitude is constantly repeating and the momentum amplitude of each item is variable; the momentum magnitude is then depending on the mass magnitude of the item transported!

Point **#9**: In regards to the Intelligent design argument: Even the Intelligent Designer would use energy to perform inertial mass motion; because, his own impulse invention is isomorphic to his centripetal acceleration invention used for the stability and radius spacing of planetary motions.

Flywheel physics demonstrates the relationship of energy to impulse. Has the flywheel energy storage been used successfully for motivating vehicles? Yes, of course. The first successful use was for a public transportation bus called the "Gyrobus" engineered by the Swiss Orlekon company, the technology is being continuously improved for energy storage systems.

Example **#3**: A further example of flowing work/kinetic energy is the large flywheel mounted directly on a DC motor-generator shaft. The mechanical/kinetic energy developed by the motor pertaining to formula #1B is flowing into and accumulating into the flywheel mass in the form of the angular velocity magnitude of the mass. When the motor-generator is switched to generator mode, the stored kinetic energy (potential kinetic energy) and the momentum contained within the flywheel is flowing back from the flywheel into the output of the generator. This action represents the creation and the conservation of momentum in energy form. This mechanical arrangement clearly demonstrates the reversible flow, the conservation and proportional relationships of kinetic energy onto mechanical energy having a flow direction, a source and a sink. This arrangement also validates the practicality of Huygens method of using formula #1, #1.1, #1B to #1.10 for mechanical machines wherein oscillations are present. Furthermore, this arrangement is also used by the presented IP device. In view of the electro- magnetic- dynamics of the DC motor -generator, is it more professional, valid, advantages or economic in thought to use electrical current flow in the time duration (the time domain analysis) instead of the root cause input energy flow? We are able to argue that the current flow is proportional to the torque delivered by motor. The torque is proportional to the acceleration of the flywheel and the voltage potential is proportional to the final angular speed of the flywheel by the cancellation of the inherent rotating vectors applying to the motor magnetic fields! No, this is not necessarily providing us economy of thought because we are then having the race horse assumption: Electrical Current alone does **NOT** describe what is making the torque follow the flywheel angular acceleration! The current flow time duration and the current flow average magnitudes are both interdependent on the energy storage capacity of the flywheel, the current supply magnitude potential, the division of the motor rotor / stator angular electromagnetic field force into distance sections of the circle and the current impedance of the motor- generator; all this is placing the motor /generator / flywheel technology firmly into Galileo's equal distance notched board displacement domain analysis. The angular speed of the flywheel * torque = energy flow, VA and **the voltage potential is the only prime root** cause having two possible variable (manipulate- able) parameters: Voltage potential and the total circuit resistance including the circuit reactance, wherein the average energy flow,

$$\mathbf{V}_{voltage} * \mathbf{A}_{amperes} = Voltage^2_{potential} / (\mathbf{I}_{mpedance, Z, electrical, total-magnitude}).$$

The system as a whole is based on the feedback principles of energy, wherein the balance of the potential energies are pinching off the current flow, like the Toilet- Tank control. The current approx. magnitude average therein is: I=Circuit voltage potential /(total Impedance Z) and the time duration to reach balance of potential energies is: t= flywheel capacity, Ws / (voltage, potential * current, VA). Accordingly, the time duration is a complex function of the flywheel moment of inertia * Impedance, which is congruent with a dampened spring oscillator. Here again is the "Vis viva" principle of formula#1. One has to consequently laudably present that kinetic energy work, VA (Volt*Ampere), **is** the **fundamental principle describing the technical potential of this system**. In short, let us say that, the kinetic energy stored within the flywheel is proportional to the monetary fuel cost of that energy; while the momentum of the flywheel is in isomorphic (diminishing returns) symmetry to the monetary energy value. This means plenty of money can be stored in a flywheel in the form of energy in relation to the motion quantity of speed. Again the "Vis Viva" is surfacing! The kinetic energy storage capacity of the flywheel is ideally suited for the temporary storage of kinetic energy because of the quadratic magnitude of the energy content in relation to the flywheels' angular velocity magnitude, angular motion and angular momentum. The ideal

Flywheel is a flywheel made of a very heavy material in respect to its volume, like Lead or Uranium, and having the stencil strength of Carbon Nano Tubes or Kevlar for withstanding the enormous centrifugal forces. Is it possible to extract every bit of kinetic energy stored into the flywheel back into the electrical energy supply connected to the generator? Of course, all physics processes are reversible, but it requires a complicated arrangement of electrical switching apparatus, which is in mechanical terms an infinite ratio progressive variable transmission, or a mechanical transmission, working with step displacements repeating in very fast cycles. Such a transmission arrangement is like sipping an espresso coffee directly from an Espresso-machine in very small quantities; a very energetic experience in very small steps, it is a machine working with quantum physics. "Is friction in any form, such as flywheel bearing friction, air friction or carriage rolling friction to be considered when describing the working physics of such a machine?". No, without friction the flywheel storage would never need recharging because kinetic energy can be recycled by regenerative breaking. Friction is consuming the energy stored within the flywheel.

Here we present **#4** example of energy flow within machines. The concept of motivating a vehicle with kinetic energy obtained from the store of angular mass momentum contained within a flywheel brings up a centrally important question: "Is kinetic energy or momentum, the product of inertial mass multiplied by velocity, a correct analysis for such a system?". Engineers will automatically resort to kinetic energy flow because the scalar magnitude of kinetic energy per time interval in Kwatt, represents the physical quantity the motor-generator delivers in the first place. Kinetic energy can be calculated into a vector impulse or momentum quantity if the need arises using the isomorphic symmetry of energy and momentum. Science courses like to concentrate on momentum, because momentum is also an important conserved physical quantity during inertial mass collisions, as demonstrated with simple physics demonstrations using the collision of carts. The sum of all the carts' momentums remains constant during their collision time interval. How many of our technologies are based on momentum conserving collisions? Why are we spending so much time on collisions?

In contrast, the very practical reason that engineers use the flywheel for the Gyrobus is the quadratic progression of the kinetic energy storage capacity in respect to the angular velocity of the flywheel, a few more very high ++3000 flywheel RPM squeezes out 50 more acceleration-trips at the so much lower bus speed limit of 50 Km/h. The Gyrobus is transferring the fuel value stored into the flywheel and transfers the fuel value through a transmission into the bus. "Is there a fictitious Force consideration applying to the flywheel stored kinetic energy?", "No, there are no fictitious forces present because of the symmetries of the inertial mass motion mechanical system; fuel values are real and only real fuel value transfers applies!"

Here is the Lagrangian-Hamiltonian **Ke** balance of this system applying to formula#1 and applying to the energy conservation of the system, no energy is lost or gained. In many publications, this balancing is actually described as balancing the mechanical books in terms energy. The pesky Cariolis "2" averaging factor is cancelling out and there is **no need to use any calculus expressions**, because the energy is transferred in tiny portions of intrinsic quantities in serial mode without loss. The integral of the energy flow is here also correctly presented by the difference of the squared velocities as presented in example #2, wherein the bus velocity origin is zero because it starts from a standstill at each bus stop $V_o=0$ and the flywheel angular speed is progression is

$$\omega_a > \omega_b: \#1.b.1) \quad I_{,flywheel}(\omega_a{}^2 - \omega_b{}^2) = mass_{,Buss}(V^2{}_{speed,limit,city} - V_o{}^2)$$

Formula #1.b.1 is closely related to the Inertial Propulsion formula #13

Then We ask: "Is it possible to sell the Gyrobus if it is engineered to deliver the momentum quantity of $p = m_{,bus}50Km/h$ gained by the bus for every starting motion from bus stop to next bus stop when this momentum quantity is withdrawn from the flywheel for every starting motion?". The scalar value of the flywheel momentum loss, in comparison to Gyrobus gained scalar momentum gain, is a grand total of only **TWO** trip accelerations!?.....

Next is the picture of the Gyrobus Flywheel assembly at Orlecon Co. 1954.

Then we must ask:

"Is the removal of momentum from the flywheel and bestowing momentum into the bus through the path of a transmission a form of collision?". "**NO**, this is not a collision because displacement distances are translated!". "Is the sum of momentums of the flywheel and the bus constant for such a large momentum differential?". "**NO**, the scalar sum of momentums at such a large momentum / impulse / velocity / torque differential is not constant within the Gyrobus system!". "Who is correct here?". The answer is obvious, because, the Gyrobus performed exactly the way the engineers calculated using kinetic energy flow. That's why the presented continuously rotational inertial propulsion works. It works with mass motion kinetic energy flow through transmissions and not direct momentum conserving collisions of inertial masses.

To illustrate again the profound difference between impulse/momentum and kinetic energy flow, let's us work out a very simplified algebraic example: Using Impulse/momentum only 2 trip start accelerations are possible:

1000(mass, flywheel)*3000(Velocity, flywheel) - 2trip*(30000(mass, buss)*50(Velocity, buss) = ZERO

When using formula #1B, #1.b.1, pertaining to kinetic energy, 50 trip start accelerations are possible. Therefore, this proves that the velocity/torque differential between the flywheel and the inertial propulsion devices' aggregate sum of masses' is too large to make it correlate to rotating vector momentum, impulse and collision, this therefore proves that:

THE CONSERVATION OF MOMENTUM, DOES NOT IN ANNY WAY, APPLY FOR MACHINES WORKING ENTIRELY IN THE DISPLACEMENT DOMAIN;

only scalar value conservation of kinetic energy applies and no calculations in regards to conservation of momentum are performed. In many publications the Gyrobus type energy conservation is defined into:

Momentum is conserved in Kinetic energy form.

This analytical view preserves the universal reach of momentum conservation and applies formula #1a; energy is the product of momentum and average speed. The Gyrobus system, when viewed as an energy flow system, is surprisingly simple: Energy flow in kwh is charged into the flywheel moment of inertia "**I**"arriving at the total stored flywheel kinetic energy =

$$\mathbf{Kwh_{totoal}=E=\tfrac{1}{2}I_{flywheel}\omega_a{}^2 = \#trip, \ starts * Ke_{,bus,speed,limit}.}$$

Accordingly: In view of the engineering reality of the Gyrobus, this publication reiterates the limitations placed on the conservation of momentum law within most good Physics books and expands the limitations with certainty by postulating: Momentum is conserved for the time duration of a direct collision impulse of point size masses. The scalar value of momentum is not conserved for the time duration of a collision of masses which have a large momentum differential and the impulse is transmitted through a complex transmission mechanism; then the magnitudes of velocity and torque are converted in relation to the transmission ratio. The momentum is then translated according to the conservation of kinetic energy law, which is the square root out of the sum of quadratic polynomials. This principle can be further postulated as: Mass motion kinetic energy transactions through transmissions are the root cause and are the prime motivating agent; while impulse magnitudes follow in an isomorphic symmetry. Accordingly, within isolated system of machines the Vis Viva, Lagrangian and Hamiltonian principles apply:

Potential Energy is primary while impulse and momentum magnitudes follow the energy transaction.

The author was unable to determine the vital rational need for postulating that momentum is insurmountable ALWAYS conserved. While we have proven with certainty that the conservation of the **transient** quantity of momentum does not apply to the Gyrobus, nor does it apply to any of the six examples when considering the reversal of physics principles, nor does it apply to the Inertial Propulsion mechanisms. However, it can be postulated, with certainty, that the sum of energies in their varied forms and the energy in its momentum form is always conserved. The herein presented straight line displacement combined with rotational motion Inertial Propulsion uses the two before mentioned vector dimensions of mass motions, the rotational and straight line mass motion. Two kinetic energy streams of these two inertial mass motions are working, side by side in an undulating energy conserving flow, inside the propulsion mechanism. Therefore, we receive one resultant reciprocal (reactive) motion of the propulsion vehicle. The kinetic energy required to motivate a body of mass is transmitted by the force impulse. In the case of the airport luggage conveyor, the tension on the belt is the force. When the tension on the belt is multiplied by the time duration of one complete belt cycle it becomes the force impulse per belt cycle time. Therefore, if we consider the conveyor with the ability to transport variable amount of mass depending on the belt friction, this publication postulate with certainty: Work/Kinetic energy flow per time interval can be mathematically extrapolated to the magnitude of a repeating force impulse applied to a defined size of mass per time interval. Therefore, this publication postulates with certainty:

A scalar Work/kinetic energy quantity generates a defined scalar impulse intensity on a defined quantity of mass by isomorphic symmetry. The scalar impulse quantity is converted into a vector Impulse by the vector geometric guidance of a mechanism.

The guidance of a mechanism is a universal property of physics which is evident in inertial mass motion, as well as in electrodynamics, thermodynamics and in radiation where diodes and mirrors can provide energy with direction.

The kinetic energy stored into the body of a mass, as the result of a force impulse, is the momentum contained within the body of mass. The momentum is the product of velocity multiplied by the body's mass.

The incremental kinetic energy content of a mass, energy gained as the result of the force impulse and expended from the store of potential energy available within the vehicle, is measured in Nm, J, Kgfm, kwh, kcalh and horse power hour. The energy quantity is in all cases the same real energy originating from the potential energy stored within the vehicle. Every reader of this publication can relate to the kwh consumed on the electric bill. "But why are we billed in kwh (energy) instead of kgfh or Nh (impulse)?". "Because an eggbeater takes four times the energy to deliver twice the rotational impulse, we are charged in energy because of the isomorphic symmetry of impulse to energy, work", then the energy to impulse relation in respect to $V_{orgin}=0$ is:

#5) Energy, work, Kgfm = impulse^2/2 * mass

Therefore: **#6) Impulse, Ns =(2 * mass * Energy, work)$^{1/2}$**

The Electricity utility would go bankrupt delivering four times the quantity in fuel and bill double the amount in Kg force hours; the impulse magnitude in relation to 1 kg mass motion. The same logic applies to the Ampere*hour=force*hour type electricity meter; we cannot envision the electricity company install this 200 year old technology because they do not respond to any potential voltage fluctuations and do not respond to the magnetic loads of motors. Despite this, there are still today the printed notion from universities fresh off the press, that impulse and momentum is the true measurement of physics, that there are "no Energy Meters" possible in Physics; a notion far beyond reality of the modern world today. Our society is based today on the economy of the measurement of energy per witched manufactured; after all, energy is measurable, expensive and worth saving; this is proven with example #1,#2,#3 and #4.

The relationship of impulse and momentum to the directional flow of kinetic energy applying to the two vector dimension of mass motion within a geometric plane is the most important aspect of the inertial propulsion and **by far**, the most often applied formula for machine design. Thereby, the very most basic principle is the end result of the inertial propulsion force impulse process which must be the transfer of a portion of the stored potential energy contained within the vehicle into one preferred direction of the whole combined mass of the vehicle. The transfer of kinetic energy into the whole isolated system of the vehicle has the result of the desired directional velocity gain of the vehicle, the resultant motion of the vehicle. Kinetic energy is the energy consumption-factor per internal self-contained impulse of the vehicle. If we now combine the formula #1a for average Energy-work with the impulse formula #6 then we arrive at the relationship of impulse to speed gain and speed averages which are each mean values of the energetic effort valid for any speed average:

Formula **#7)** $\mathbf{P=Impulse,Ns =mass\ (2 * Speed_{gain} * Speed_{average})^{1/2}}$

$Speed_{average}=s/t=Speed_{gain}/2$; $V_0 = 0$;$s=$ cyclic repeating ; t is cyclic variable

$$\mathbf{P=Impulse,Ns =mass\ (2 * Speed_{gain} * s/t)^{1/2}}$$

Wherein $Speed_{gain}$ is also expressed with dV in calculus courses or with: $V-V_0$.

Formula **#7** indicates that the total impulse is the diminishing returns relationship of the **average speed**, when the cyclic repeating displacement speed gain amplitude and displacement **s** is in-variable repeating. The V_{orgin} is zero within formula #7. From here, it is already possible to extrapolate the proportional relationship of spin angular frequency (a measure of rotational distance per cycle time) to impulse and that a self-contained impulse within a uniform repeating displacement length magnitude **s** reciprocal straight-line cycling system might be possible. We have also seen from the airport conveyor example #1 it is not possible. "Do we have a paradox because of inconsistence analysis?". The incongruence has to do with the 2 modifier in formula #7. The above formula is guaranteed to deliver the true (net) effective impulse only if the speed average is 1/2 of the speed gain amplitude which is then applying to a uniform progression straight line displacement motion. This is why we find the statement: "Only applicable to uniform straight line displacement motion progression," all over Physics Books. However, the impulse magnitude returned by formula #7 is less than what is being measured with load sensors, digital integrator and a scope within a rotational to straight line displacement inertial mass motion. This is because the impulse returned by #7 which employs the square root out of 2 modifier, the root mean square, must be regarded as the minimum real (net) effective impulse magnitude for perfect straight line motion without rotational coupled motions. We must further analyze what Newton meant with his statement, "too tedious to analyze all possible combinations of rotational to straight line displacement coupled motions reflections".

Here we arrive at the point where most of the fundamental principles have been discussed. We now proceed to the prime principle of inertial propulsion. However, the primary inertial propulsion principle is also an example of potential energy distribution and must be view as example #5.

The **primary** IP principle is the distribution of an initial condition root cause potential mechanical energy between two unequal bodies of mass having a simultaneous, mutually reciprocally unimpeded separating motion which is caused by the power of one single source of potential mechanical energy. The whole assembly of all the parts of the vehicle is the larger mass; the straight-line (cyclic back and forth) moving inertia element (the flywheel-rotor assembly) within the vehicle is the smaller mass.

To begin with: There are two energy distribution motions and two energy collecting motions having unequal initial potential energy states within one complete IP cycle, applying to the combined rotational to straight-line displacement coupled motions reflections (translation). The impulse is, accordingly, a difference of average velocities and regular repeating base velocity amplitudes applying to formula #7.

Example **#5**, Is, by far, the most important and **complex** kinetic energy machine applying to inertial propulsion! Here we consider: Two bodies of **UN**EQUAL inertial mass are simultaneously mutually and reciprocally separating by the force of one single compression spring and are guided by a frictionless mechanical arrangement in one opposing vector direction of motion.

WHAT is the RATIO of the kinetic energy bestowed onto each inertial mass at the end of the accelerated separation?

This question has five (**5**) unknown parameters **magnitudes**:

Unknown1): The **magnitudes** of the **velocity gain** of the first small mass **m**.

Unknown2): The **magnitudes** of the **velocity gain** of the second large mass **M**.

Unknown3): The time duration **magnitude t** of the reciprocal acceleration.

Unknown4): The accelerated distance **magnitude l_m** of the first small mass.

Unknown5): The accelerated distance **magnitude L_M** of the second Large mass.

To visualise the vector-kinematic concept of the separation of two unequal masses from a single source of mechanical energy a picture is provided; this concept was successfully proven by the Author on April 3rd 2008 and it exists only as a calculation procedures in our science and not as a formal theory:

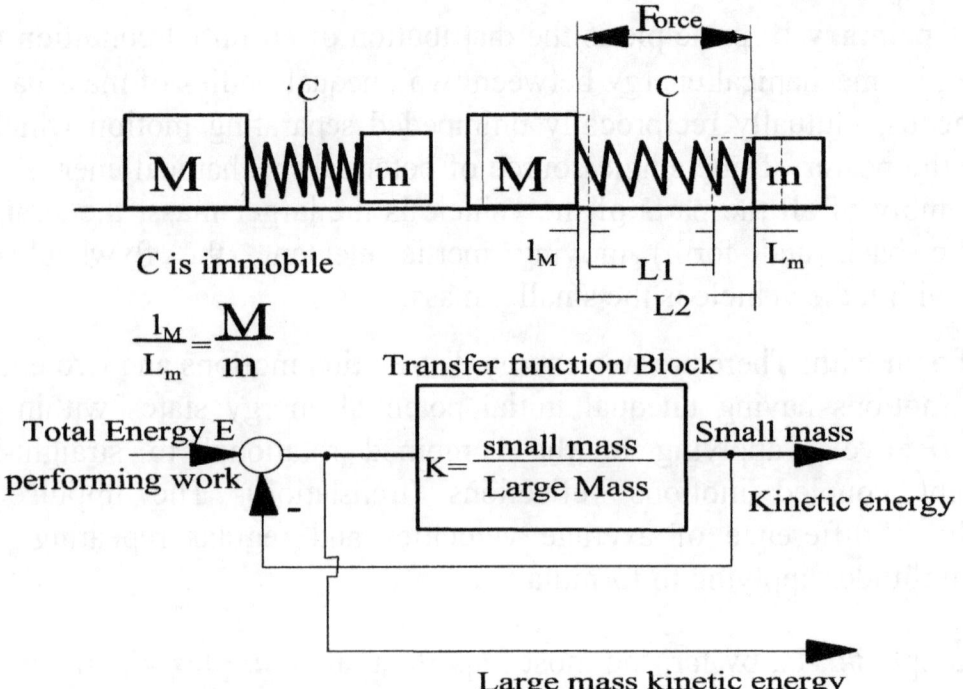

C is immobile

$$\frac{l_M}{L_m} = \frac{M}{m}$$

Transfer function Block

Total Energy E performing work

$$K = \frac{\text{small mass}}{\text{Large Mass}}$$

Small mass

Kinetic energy

Large mass kinetic energy

We know from Newton that impulse, the product of the spring force contact TIME multiplied by the force magnitude MUST be equally applied to each body of mass; but we don't know the contact **time duration**. Therefore, we do not know the **MAGNITUDE** of EQUAL reciprocal MOMENTUM of the two masses derived from one single source of potential mechanical energy and thereby the kinetic energy distribution RATIO, because we **do not know the time duration** of the force applied nor the velocities of each mass nor each individual acceleration distance?

The one single first original root cause potential mechanical energy is distributed into two kinetic energy magnitudes by a dynamic distribution RATIO of:

THE INVERSE RATIO OF THE SEPARATING MASSES.

In algebraic form:

$$\text{Energy, kinetic, large, MASS} / \text{energy, kinetic, small, mass} = \text{mass, small} / \text{Mass, large}$$

The smaller mass receives the larger amount of kinetic energy.

Formula **#8)** $$E_{large}/e_{small} = m_{small}/M_{large}$$

Formula **#8** also proofs that the V_{origin} magnitude **is in-material** in relation to this distribution of energy principle; the V_{origin} can be zero or at the speed of light the distribution of energies still holds true because the impulse time duration is based on the reciprocal velocity relation between mass **M** and **m,** **and** is zero at the start of the reciprocal motion. Accordingly, If IP is proven then also the unlimited speed limit of IP is proven. The time duration is also a function of the inertial mass magnitudes: $t^2 = 2l_M M/F$; $t^2 = 2L_m m/F$; $t = (2l_M M/F)^{1/2}$; $t = (2L_m m/F)^{1/2}$

Then the reciprocal expansion length of the spring is depending on the mass magnitude:

$$l_M M = L_m m$$

Proof:

The total energy of the system is constant and accordingly is:

$$Energy_{total} = Energy_{kinetic,large,Mass} + energy_{kinetic,small,mass}$$

Therefore: By combining all three formulas we arrive at formula

#8.a) $Energy_{,kinetic,small,mass} = Energy_{total} / ((mass_{,small}/Mass_{,Large}) + 1)$

Formula #8.a   is the fundamental feedback system formula applicable to all reluctance type phenomena; as illustrated in the picture,
wherein the ratio of the separating masses is the open loop transfer function:

k = m/M.

Furthermore:
The product of mass and kinetic energy is equal for each separating mass.
The product of kinetic energy and mass must be viewed as mechanical kinetic energy momentum of mass.
The Mechanical Kinetic Energy Momentum gain is equal for the separating masses.
By introducing the proven definition of kinetic energy $= E = \frac{1}{2} m * V^2$, starting from equal momentum in algebraic form: $Mdv = mdV$; $M > m$ and $dV > dv$
Now squaring both sides gets: $M^2 dv^2 = m^2 dV^2$
Breaking out the ratio of the masse we get the energy distribution relation:

$M/m = mdV^2/Mdv^2$

The product of mass and velocity is equal for each separating mass, which is Newton's momentum;

$$Force_{,average} = mass * a_{ccceleration}$$

The product of mass and velocity is equal to the product of Force in the time duration, which is IMPULSE? And further:
the product of mass and acceleration is equal for each separating mass;
the product of mass and acceleration is average Force;
the Force is equally applied onto each mass;

the total center of the combined mass, the CM, is stationary in relation to the opposing motions; therefore we have proven that $M/m = mdV^2/Mdv^2$ is the third Law expressed in the energy potency form.

For Validity:**Ref. Schaum, 3000** solved problems in Physics:**Problem 4.15**

Important 7 Points to consider:

Point **#1:** Formula **#8, #8a** is Newton's third law in the displacement domain **feedback energy potency form**, it is expressed in a vector-kinematic feedback loop **proof** picture at the beginning of the sample **#5**; this is a formal proof of this derivation and **it is the only form** providing the actual reciprocal impulse and kinetic energy magnitudes from the mechanical energy first original root cause potentials. The SRT derivation, in contrast, has no formal feedback loop vector-kinematic derivation to support it at this time, even thou it has the feedback term: $(1-V^2/c^2)$ in the denominator. The formula #8, 8a Theorem holds true for uniform force and for a non-uniform force applying to each displacement distance increments **ds**, it also holds true for SRT and none SRT arguments; this is because this holds true for every minuscule delta time **dt** and delta distance **dx** of the mutual separation because the mass ratio never changes, because the V_0 is Zero in relation to the mutual and reciprocal separation and the **mass ratio** is the **dynamic governing relation** here! Accordingly, the pesky 2 divisor for the uniform motion consideration of formula #7 has again been eliminated!

Point **#2:** The Theorem holds true at any Vo speed of the moving inertial frame, even beyond the speed of light, because of Einstein: The physics of a moving platform is invariable, even at the speed of light and beyond. Accordingly:

Formula #8 and 8.a is the basis of inertial propulsion!

Point **#3:** This is a very important repeated fundamental principle in Physics which must be further expanded to a compound feedback system for the presented inertial drive system wherein an internal straight line displaceable flywheel axis motion is used.

Point **#4:** The author is unable to determine as to whom or as to when the mechanical to kinetic energy distribution ratio was discovered or first used. Newton did not use the term ENERGY or the play of forces in the displacement domain until his pendulum derivation in his last Principia book. Newton was satisfied that the impulses were equal and the two momentums were equal; no **question** was raise by Newton as to the actual **momentum, vector velocity and time Magnitudes,** these questions were first raised by continental scientist Huygens, Leibniz, Bernoulli and British scientist Hooke. We don't know how Newton would have solved the **momentum / time magnitude problem** with his equal impulse/momentum laws because momentum only analyses the <u>change in momentum</u> independent to the velocity magnitude, while the formal kinetic energy-work theorem **does** analyse the velocity magnitudes. The mutual reciprocal concept could be extrapolated from Huygens' "Oscillatorium" paper and is taught always in calculations when the first original root cause of an inertial mass motion is a potential mechanical energy source.

Point **#5:** Importantly, the potential mechanical first original root cause energy source can equally be a compressed spring or a spinning flywheel which supplies mechanical energy through a transmission. Then the need arises to correlate the potential mechanical energy of the flywheel to the resultant impulse.

Point **#6:** We have to ask: "Why is the energy distribution feedback flow ratio concept not included in our formal physics books?". "Why do we learn these relationships through sample problems in the "Schaum Books" series instead of a formal stated law?". "Why do we have to first use Newton's equal impulse / momentum relationship and then expand the impulse to mechanical energy momentum using the $V_{average}=S_{distance}/t_{time}$ donkey bridge?". In reality it is a mechanical energy distribution feedback relationship in the first place and **we have proven that energy is the first original root cause of the reciprocal impulses**. It was in fact invented **before** equal reciprocal impulse; it has the easy Lagrangian, Hamiltonian and Hertz **technology** compliant formulation; this makes the Example #5 mechanics in reality an **Energy Transducer, it is Newton's third law in the feedback energy form!** To visually show the third law in energy form a mutual reciprocal motion of a large flywheel mounted onto an electrical motor housing and a small flywheel mounted onto the output shaft, the ratios of square of the angular speeds will be in reverse to the ratio of the moment of inertias:

$$I_{large}/I_{small}=I_{small}V^2_{small}/I_{large}V^2_{large}.$$
Please view **youtube, ggutsch1,P1100007.**

Point **#7:** Are there more important machine constructs using the Third Law in energy form, the distribution of mechanical energy in reverse relation to the inertial mass ratio? Yes, there are many examples; one noteworthy example is the first, most simple and effective sub machine gun uses the Third Law in Energy Form by mutual, reciprocally and simultaneously accelerate the **bullet** and reciprocally the breach lock-cartridge **extractor bolt**, whereby the bullet will travel to the end of the 10 inch gun barrel while at the same time instant the relative massive breach-lock bolt mass travels in an opposing reaction distance only 1/8 of an inch according to the ratio of bullet mass to the large bolt-mass and still maintains enough kinetic energy to extract-pull out the spend cartridge and load the new bullet in an endless automatic cycle. This small 1/8 inch motion of the breach-lock-bolt during the firing of the bullet allows the retention of the explosion forces against the cartridge walls within the breach, thus preventing the cartridge to be deformed or jammed. The bullet mass to breach extractor bolt mass ratio has to be carefully fined tuned using formula #8, #8a in congruence with the IP fine tuning of the impulse flywheel (1) angular moment of inertia to the straight line motion inertial mass magnitude tuning presented in next examples #6 and #7; this places the IP machine into the very complex category only fully engineer-able with numeric computer algorithms presented by the author further on!

Example #6: We now will consider the special case of the mutual and reciprocal separation between an inertial mass in a straight line motion and a rotational flywheel on a **fixed axis**. Important point, inertial propulsion is performed by a **movable flywheel axis in a straight line motion**.

The motive power is provided by the stored mechanical energy of a flywheel. "What is the distribution of the mechanical energy?".

The ratio of the rotational moment of inertia to the straight line inertial mass times the squared motion radius is the reverse ratio of the kinetic energies impressed onto each part:

#8.b) $$m\ r^2/I_{flywheel} = e_{rotational}/e_{straightline}$$

For Validity Proof **Ref. UCSD department of physics course web pages**.

Here we present Example **#7** on the next page, the fundamental principle of the continuous spinning Inertial Propulsion, wherein the realisation of this principle into a workable IP product becomes a technology challenge and not a science impossibility.

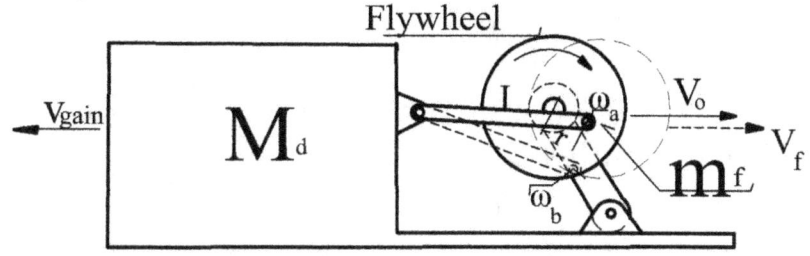

$$I = 1/2\, m_f\, r^2$$

Initial condition potential kinetic energy $= 1/2 * I * \omega_a^2$; $V_o = 0$

Distribution ratio of energies:

$$\frac{M_d}{m_f} = \frac{m_f * V_f^2}{M_d * V_{gain}^2}$$

Total Energy distributed $= 1/2\, I\, (\omega_a^2 - \omega_b^2)$

ω is dependant on the gravitational pull and ratio of masses

and is obtained with the mutual and reciprocal rotational exertion against a second flywheel, as presented in the next skater and trebuchet examples where two rotors are employed on the same axis of rotation; the dynamic exertions between two mutually opposing rotating rotors on a movable axis is the **ESSENCE** of the Author's patented inertial propulsion machines. Dynamic exertions means: energy can be delivered or absorbed mutually and reciprocally between the two opposing rotors, this is the dynamic breaking in cars.

The whole aggregate mass M_d is being propelled by the gain in velocity V_d independent of the previous velocity origin V_o; the previous velocity magnitude is not a part of Newton's $V_d = p/M_d$ derivation applicable in this example. According to Newton, V_d repeats on and on forever independent of the velocity magnitude. This is the "Accordion player motion or simulates mass expulsion motion" argument presented for formula #1.5. Then this publication postulates with emphatic certainty:

The dynamic principles of mutual separation of unequal inertial masses and flywheel Physics, the distribution flow of mechanical energy on the bases of the reverse ratio of the inertial mass motion magnitudes within a feedback loop, it is a mass motion Physics Principle standing on its own, it is a **far reaching principle**. It is, in fact, Newton's unfinished theorem. This principle is reversible, when we change the flow direction from momentum into a mechanical energy storage device, we recapture kinetic energy into mechanical energy; then momentum is conserved in energy form. This is applying to the reversibility of formula #6 and #8.

Accordingly, we have proven that within all the presented seven machine examples Newton's momentum is conserved in energy form.

While Huygens "Oscillatorium" paper was still largely based on geometric vector-kinematic constructs, it provided displacement based analysis shortcuts to solve the pendulum problems of clock escapements which are not directly taught in to-days Physics books.

The kinetic energy distribution ratio has the consequence that the body with double the mass receives **1/3** (which is less) of the total potential energy of the compressed spring system and the body with ½ the mass will receive **2/3** (which is more) of the total potential energy of the system. That the energy distribution process is a feedback system should come as no surprise, as so many systems are feedback systems, from H. Hertz's electrodynamics and Darwin's Biology to the collapse of the stock market. All these are working with feedback systems. IMPORTANTLY!! Kinetic energy, however, was discovered by Huygens / Leibniz and its importance rediscovered 100 years later by Lord Kelvin. The example of solving the separation of two unequal bodies of mass, separated by one single source of potential mechanical energy is a displacement domain analysis; the play of forces in respect to the displacement of the masses and it is in symmetry with Quantum Mechanics. Then we can postulate:

Within rotational to straight line motion coupled (translated, projected) **mechanics the original root cause of motion is <u>always mechanical, electrical, chemical or nuclear energy</u>, while momentum is only temporarily (a very short delta time) identified from the undulating energy flow and does not represent a physical or mathematical factor at all; therefore, it is incongruent to assign the conservation of momentum as a limiting factor for these class of mechanics.**

In Physics text books the subject of mutual separation of bodies is invariable presented as a push between two **equal sized** Ice Skaters. Then, according to Newton, there is an equal reaction to an action of equal momentum and also of equal reciprocal velocities. So, now let us discuss an example of two groups of skaters skating in opposing circles an example never included in physics books. Next are two machines with self-contained single Inertial Propulsion impulses; these machine proof the limited scope of the conservation of momentum conservation and that it **is not preventing** Inertial Propulsion.

Does the conservation of momentum have a universal reach to prevent the performance of one single Inertial Propulsion impulse by a group of Ice Skaters?

Let us now consider a matched group of 3 skaters and a group of 2 skaters wherein each group has equal mass. The group of three skaters are matched so that one large (fat) skater has equal mass in respect to two opposing smaller skaters in orbital motion. The group of two skaters each have an equal mass and their combined mass is equal to the first group large skater plus one small skater. Each group using a ridged tether for help in maintain skating in opposing pre-determined ex-centric circles marked onto the ice surface, wherein the group of two skaters will provide the **inertial backrest** of the eventual self-contained impulse and have a larger circle and are taller to clear the group of three skaters. If we cause a mutual energy absorbing collision between the two circulating groups, such that we anticipate the shifting of the group of three Center of Mass being equal to the circles centre of mass ex-centricity when only one small skater of the group of three slides off free with his acquired rotational tangential vector. Then we will find, **it is possible**, that 2 minus 2 skater collided kinetic energy, angular momentum and centrifugal forces are absorbed- collapsed to zero. This collapse is the disappearance of momentum, the conserved quantity of momentum disappearing under the influence of energy flow; this is possible because of the conversion of momentum to energy a process involving the square root out of $(-1)^{\frac{1}{2}}$, a math-function returning no valid result according to Newton's era math understanding, the $(-1)^{\frac{1}{2}}$ math function in relation to rotational dynamics was solved much later by a mathematician named Gauss. The Exception is that the one free skater tangential vector momentum is unimpeded, is conserved and causes a **self- contained impulse** against the rink boundary! The issue of a rest torque being applied onto the rink ice is cleared by a second mirror image simultaneous experiment; a principle postulated by Prof. Eric Lathwait. The principle of disappearance of angular momentum with a mutual reciprocal energy absorbing collision is a concept not available in "Newtonian Mechanics"; in Newtonian Mechanics angular momentum is always conserved, it is indestructible! The issue of dynamic energy absorption (dynamic breaking action) is well solved with energy absorbing materials; for example, soft bendable tin wires or rubber band technology having an exact stress break point. Such a breakable rubber band or rubber cushion must have a permanent deformation, wherein no snap-back or recoil-reaction occurs; the rubber / spring deformation must be permanently freezing the captured energy into the material. This removal of momentum and storing it into the material of a spring / rubber deformation is unknown in Newtonian mechanics; this is the only reason why inertial propulsion is inconsistent with traditional Science, but it exists in nature why not accept it? **Then we can postulate:**

Inertial propulsion is <u>singularly</u> a technological problem of energy management between two in opposing direction rotating bodies on a common axis and it is therefore <u>proven</u> to be possible.

The use of angular mutual and reciprocal simultaneous exertions between **two opposing** rotating bodies on **a common rotational axis employing dynamic energy absorption** resulting into a straight line self-contained impulse against the centre of mass of the same system itself is the **ESSENCE** of the author's patented inertial propulsion machines; wherein the mutual and reciprocal principle is example #5,#6,#7 and formula **#8, #8a. Here is a picture of the Ice Skaters:**

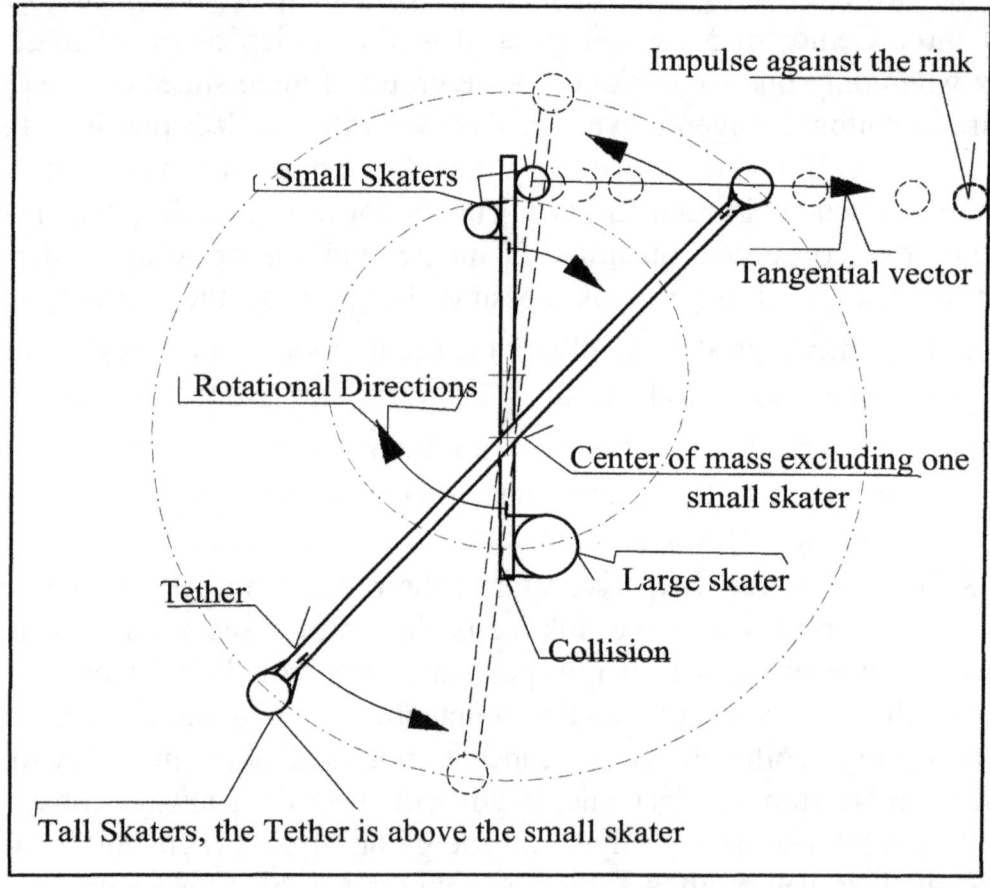

Impulse against the rink

Small Skaters

Tangential vector

Rotational Directions

Center of mass excluding one small skater

Large skater

Tether

Collision

Tall Skaters, the Tether is above the small skater

With the Ice-Skaters example we have presented the first simple example of how inertial propulsion is physically possible! It is Newton's unfinished Theorem. For such a rotational system involving collisions we must "<u>discard</u>" the notion of stick-on, friction and reiterations as the root

cause of IP <u>and</u> we also must discard the disparaging notion that IP is related, in any way, to claims of free energy devices!

Presenting a Machine for exerting one single Inertial Propulsion Impulse.

The earliest example of using the combined vector sum effort of straight line displacement and rotational kinetic energy to produce a large straight line displacement force impulse is the carriage mounted medieval catapult called "Trebuchet". The carriage of the Trebuchet is not only used for positioning but its main function is to improve the projectile range. The improvement in range of this catapult was apparently due to the simultaneous combined effort of straight line motion kinetic energy, rotational kinetic energy and the time spaced delayed lever action of the whip attached to the throw arm. This delayed action is profoundly different than the instantaneous action of impulse and velocity gain in a straight line inertial mass motion. Since Newton pronounced, as presented on page 17, to keep rotational motion projected onto straight line displacement reflections out of his Principia; the trebuchet principle cannot be found within it. However, the centripetal acceleration which gave us the understanding of planetary arc motion can be found in the Principia. Newton regarded the centripetal acceleration as a separate invention from the straight line displacement mass motions, because it is exclusively an energy transaction. The centripetal / centrifugal acceleration is also the underlying principle and the root cause of the remarkable trebuchet operation. The straight line displacement motion component contained within the total trebuchet motion, which is operating in the direction of the throw of the projectile, is caused by the large inertial reluctance of the counter weight reciprocally inducing a straight line displacement motion into the carriage and also into the throw-arm from the rotational potential energy of the counter weight, a reciprocal motion applying to formula#8.

Any carriage friction to the ground will diminish the total angular acceleration of the throw arm; therefore, friction is diminishing the range performance of the wheeled trebuchet. Accordingly, friction is not the first root cause of the trebuchet operation as claimed by the stick-on / friction arguments contrary to inertial propulsion and must be **expelled / denied** / pronounced **redundant** from the root cause analysis procedures.

The "Trebuchet" was also the first device to generate such a large straight line displacement force by angular acceleration of a rotational rotor mass within less than one half revolution of the rotational motion employing the proportional relationship of the centripetal force over a rotational distance to the kinetic energy of a rotating mass. The angular acceleration of the trebuchet throw arm structure does not induce a net straight line motion of the center of mass CM of the aggregate trebuchet system; the CM stays stationary during the acceleration. This means: The trebuchet mechanical oscillator receives its frequency modulation independent of an external reference point, independent of a backrest or steady-rest. With fine tuning of the Trebuchet lever actions it is possible to convert up to 65% of the potential energy of the counter weight into motion energy of the projectile, depending on the transmission ratio of the whip length to the throw arm length. The lost energy due to recoil actions is only 35% of the root cause potential energy. This fine tuning application demonstrates that the Trebuchet recoil action is a variable parameter; unlike Newton's third law invariable instantaneous equal re-action to an action, it demonstrates that the centripetal acceleration has different capabilities then the purely singular straight line displacement motion acceleration. While the original carriage mounted trebuchet has only one charge of potential energy per operating cycle, the present described modified trebuchet has two alternating energy charges per operating cycle. To be fully congruent with the operation of the presented Inertial Propulsion Device, an additional mechanical pull mechanism, having an opposing potential energy charge, must be present on the throw arm to motivate an additional flywheel. The additional flywheel is independently rotatable and is mounted onto the exact final center point of gyration of the throw arm. The additional flywheel is called the counter rotating arm in the trebuchet education device picture. The additional pull mechanism is used to motivate the additional flywheel (counter rotating arm) up to an exact rotational momentum magnitude but in opposing rotational direction of the throw arm and its acceleration has no effect on the straight line momentum of the trebuchet; it is Newton's equal and opposing action. The momentum magnitude induced into the additional flywheel is as large as remains in the counter-weight/throw-arm after the throw. The additional counter rotating flywheel momentum must engage with the flywheel in a timed rotational reciprocal energy absorbing collision to oppose the counter weight/arm rest momentum remaining after the projectile throw. The additional flywheel momentum is negating the recoil of the throw arm to an exact zero momentum. Zero trebuchet-system recoil, because of the collapse of kinetic energy being the root cause of the action, then causes the reactive momentum in the counterweight to collapse as well. This root cause and effect sequence is the same as in the Gyrobus, where the rotating vector momentum quantity is absorbed into kinetic energy. Such an improved Trebuchet exerts a large, self-contained, real effective impulse component against the projectile.

Here, Newton's third law is dynamically avoided by the dual time spaced reciprocal distribution of energies inverse proportional to the ratio of the component masses; the principle applying to formula #8, #8a which is Newton's third law in the energy form.

The physics principle of the modified Trebuchet can be further viewed in congruence with the non-harmonic oscillation of an oscillator when pumped by alternating energy pulses which have a non-resonant frequency to such a power magnitude that the oscillator's oscillations are stopped, therefore have the largest possible gradient complex plane force-velocity projection. This principle is also known as forced oscillations.

The simultaneous combined straight line and rotational motion of the improved Trebuchet has non-harmonic motion similarities to the presented continuous cycling Inertial Propulsion. The improved trebuchet projectile exerts an impulse against the aggregate mass of the vehicle, while the carriage is displacing in a straight line within the vehicle. Furthermore, within the present Inertial Propulsion Device the alternating energy pulses and the mechanical leverages are fine tuned to accomplish a continuing rotor rotation having inertial mass motions with the largest possible gradient complex geometric plane projections and the largest possible average angular speeds with identical straight line carriage speed amplitudes; these parameter relations provide the optimum real effective thrust yield. It is important to note that all these relationships are rotating vector quantities from the complex rotation of the fulcrum arm structure having a direct proportional relationship with the centripetal forces over the rotational displacement distance. At the same time, the counterweight has a straight line displacement motion vector mutually opposed to the direction of the throw motion of the projectile and opposed to the carriage motion. This must be viewed as **only a potential** equal straight line motion reaction to the projectile throw action. This potential matures into a real reactive straight line motion opposed to the projectile throw action in the form of a trebuchet aggregate mass motion at the end of the trebuchet phase cycle delay. Newton's third law: "Equal Reaction to an Action" is here delayed by a 90° rotation!

While we do teach the trebuchet principle in our high schools, but "do we teach the rotating vectors phase delays of the trebuchet vector quantities vigorously enough?". "Do we teach that this phase time delay is an opportunity to reduce the potential of the reaction forces to zero with a kinetic energy transaction?" This is a reduction of momentum in the form of an energy absorbing collision.

Here **is,** where we arrive at the **exact point** where we reduce the equal reaction forces to zero without negative consequences, we actually make Newton's **equal momentum reaction to an impulse action disappear**; because of the dual existence of the straight line displacement momentum vectors also existing as **rotating kinetic energy magnitudes** at the exact time the projectile separates from the throw arm. If we diminish the rotating vector energy quantity with a **PERFECT RECIPROCAL ENERY ABSORBING** collision, we will also diminish the unwanted counter weight straight line displacement momentum vectors to zero, **we therefore prove at this point the generation of a reaction less projectile throw.** This means that we are forcing the momentum to follow the energy transaction; a complex math function involving imaginary numbers in form of the square root out of -1 is $(-1)^{\frac{1}{2}}$. Accordingly, the collision is effectively locking the projectile momentum into the whole aggregate system mass.

So here we have done IT! That's how it's done! **This mutual and reciprocal energy absorbing collision between two opposing rotors is the ESSENCE** of the present patented inventions; this is the dynamic management of two opposing Kinetic Energy magnitudes contained within two opposing inertial rotor motions independent of the straight line motion of their common **CM** axis of rotation allowing a time delayed self-contained straight line impulse exertion. Dynamic energy management means: either an reciprocal energy absorbing collisions or a reciprocal dynamic breaking actions; wherein "dynamic" means the energy distribution of the kinetic energy of example #5, it is a function of the mass magnitudes, the large mass makes a small contribution and the small mass makes a large contribution feed into a single energy accumulator. It is important to note that it must be a rotational reciprocal energy absorbing collision type where the center point of rotation must be exactly **AT** the final center of mass (CM) of the final rotating mass configuration excluding the projectile, because the projectile has separated at this point. Any offset to the final CM will diminish the null action-result of this collision leaving an unwanted rest reaction because of the quadratic factor of the Huygens-Steiner theorem. This principle should come as no surprise as it is consistent with the three energy flow examples. It is also consistent with the reversibility of physics functions where the rotational acceleration has the same principle as the rotational de-acceleration, but has different directions. The reciprocal nature of the collision satisfies the conservation of rotating momentums when we include the sum of all momentums, including the projectile momentum and the isolated systems' straight line momentum remain constant. Here is again, no issue of violation of the conservation principles occurring here, as despairingly assigned to Inertial Propulsion by traditional science.

The rotational centripetal forces are created from a force couple on the balance beam leaver connecting the counter weight and the throw-arm from a store of potential energy. The force couple are two equal forces pointing in the same direction, which squeeze against each other. The forces of the trebuchet force couple are separated by the distance between the mass-center of the counterweight to the pivot of the throw-arm, which creates a rotational moment. The force couple of the trebuchet creates an additional mass motion kinetic energy exchange. This is different than Newton's instant force couple of the third law. Thereby, the working principle of the carriage mounted trebuchet and for rotational to straight line displacement coupled motion in general,
The Huygens- Steiner Theorem applies.

The Huygens-Steiner Theorem is also called The Parallel Axis Theorem. The parallel axis theorem tells us that a forced point of rotation, which has a distance to the natural center point of gyration at the center of mass **CM**, forces a rotating mass structure to have a larger reluctance to rotation. The reluctance to rotation is called the "Moment of Inertia **I**"; therefore, the structure is having a larger ability to store kinetic energy by the square in the offset distance. The Huygens Seiner Theorem is:

$$\mathbf{I}_{offset} = \mathbf{I}_{CM,origin} + mass_{sum,total} * S^2_{pivot,offset,distance}$$

This principle has its root cause in Huygens centrifugal acceleration of rotating masses; this is because, the origin moment of inertia $\mathbf{I}_{origin}$ and the offset moment of inertia are quadratic functions of distance, when they are added we get a sum of squared function. However both $\mathbf{I}_{offset}$ and $\mathbf{I}_{origin}$ are proportional to the Vis viva, both are proportional to the Ke content of the rotating structure this very important exponential principle is very often overlooked or should we say it is simply ignored leading to incorrect derivations and conclusions. The centripetal acceleration is:

Formula #9) $\boldsymbol{\alpha}$=**acceleration,** $_{centripetal}$=$V^2_{tangential}$/$radius_{,gyration}$.

In the case of the carriage mounted trebuchet the offset distance is large because the natural center point of rotation is very close to the center of mass of the very large counter weight. In contrast, at the start of the throw arm rotational motion, it is forced to start to rotate at the fulcrum pivot. As the throw arm rotation gains speed, the center point of rotation shifts to the natural center point of rotation which is at the CM of mass of the whole rotating structure. The shifting is caused by the centrifugal force based on the quadratic formula #9 pulling at the component masses, a self centering process. This is increasing the rotational speed a great amount, because the total system energy is conserved and remains constant. However, the ability to store energy decreases with the rotational speed. This shifting from the forced center point of rotation at the fulcrum pivot is shifting to the natural center point of mass which is the preferred center point of rotation and is caused by the quadratic functional nature of Newton's centripetal acceleration applying to the component masses which cause large forces within the throw arm structure and has the force root origins at the center of mass of the aggregate rotating structure. This principle delivers a large avalanche of kinetic energy into the trebuchet throw arm tip containing the projectile. This is a similar action as seen in electro dynamic coils generating large energy sparks, as predicted by Heinrich Hertz in his before mentioned written works.

"Is it possible to use the carriage mounted trebuchet principle to power one single cycle of inertial propulsion?". "Yes, it is". The author, Gottfried Gutsche, has patented an Education Device for demonstrating the generation of 1 Newton second (one) magnitude of self-contained impulse.

The energy required to perform the 1 Newton second impulse against 1Kg mass is the energy of 0.5 kg-force meter. This is the energy of a spring compressed by 1Kg weight over a distance of ½ meter.

The progression of the operational cycle of the Education Device is shown on the "Mindbites.com" web site:

www.mindbites.com/series/1278/ lesson 7

Then we must ask, "Where is Archimedes fixed point, the inertial backrest, the steady-rest, in this dynamic process?". "What is performing the throw of the projectile within the modified trebuchet **without any recoil**?" The answer is contained in all the previous examples, but primarily, the time-delay of the potential to the final straight line motion magnitudes is the "corpus delicti". This in combination of the kinetic energy absorbing collision is the most potent explanation combination.

The trebuchet throw arm, with the counterweight and the projectile still present performs a rotation around the natural center point of rotation, just as the flywheel in the Gyrobus has a natural center point of rotation. Furthermore we can say, "a gyro is self centering by the influence of the kinetic energy content which is proportional to the centrifugal force; a process used to balance car tires". A car tire natural center of mass **must have** congruence with the center of its axle! If the kinetic energy of the trebuchet throw arm structure is removed in the same logical way as seen in the Gyrobus technology and this energy removal is the root cause of a rotational energy collapse and the rotational energy removal is at the new preferred center point of rotation, **excluding the projectile mass**, then, the rotational vector momentums contained in the throw arm mass and counter weight mass components are removed. The EXCEPTION is that the kinetic energy of the projectile is conserved which compels the projectile to fly onto the boundary of the system using the kinetic energy quantity as a momentum vector magnitude.

With the Gyrobus, the skater groups and the Inertial Propulsion Education Device as physical proof we can postulate with certainty:

A mechanism is capable of inducing a straight line directional internal (vector) impulse into a self-contained quantity of mass from a quantity of potential energy, independently of any recoil action, **ONLY, IF** the root cause of the internal mass motion action is a reciprocal rotational kinetic energy flow exchange which causes a rotational kinetic energy collapse. In view of the presented principles it is incongruent to elevate impulse and momentum as the superior elements of mass motion. Instead, the root cause of potential energy and the resultant effect of the progression sequence is the underlying principle.

Formula#6 is modified with formula #9 for converting rotational kinetic energy into impulse:

$$\textbf{Impulse, self-contained, Ns} = \left(\textbf{2mass}_{\textbf{projectile}} \tfrac{1}{2}\textbf{r}^2\boldsymbol{\omega}^2 {}_{\textbf{angular, speed, max}}\right)^{\tfrac{1}{2}},$$
$$\text{Then:}$$

$$\textbf{p}_{\textbf{second,Newton s}} = \textbf{mass}_{\textbf{projectile}}\textbf{r}\boldsymbol{\omega}_{\textbf{,speed, max}}$$

Next a drawing of the Education Device is presented for depicting the acceleration of the throw arm structure and the subsequent transfer of the impulse into the isolated system. This action is indicated with dashed outlines.

Within the drawing the identification of the motion components in relation to the following photograph of the pendulum test is as follows: The throw-arm is ///, the transfer mass (projectile) is ← ← and the counter rotating arm is +++.

The natural stationary center line at the center of mass is indicated and the forced center line of rotation is drawn at the hinge pivot. The counter weight is dimensioned to accomplish the motion of the carriage in the direction of the throw without adding excess mass to the device.

The main spring is the primary energy source of the throw arm acceleration magnitude over the rotational distance applying to formula #1c.

The mass motion vectors are indicated with arrows. It is important to point to the absence of any force vector arrow in the reaction direction of the trebuchet motion. This is because

there are no Newtonian reaction forces

in opposing direction of the trebuchet motion; they have been selectively absorbed by the energy absorbing collision before they mature into an equal straight line reaction momentum. Here we have a mixture of energy transactions and momentum magnitude results, this might be inconsistent with pure Newtonian mechanics, this is however a proven reality within machines. Then, again, we can postulate: Within machines Newton's momentum is conserved in energy form.

However, there are torque force vectors against the surface of the earth, first during the rotational acceleration and then during the self-contained impulse against the whole isolated system.

On the next page is the illustration of the modified trebuchet:

Education Device

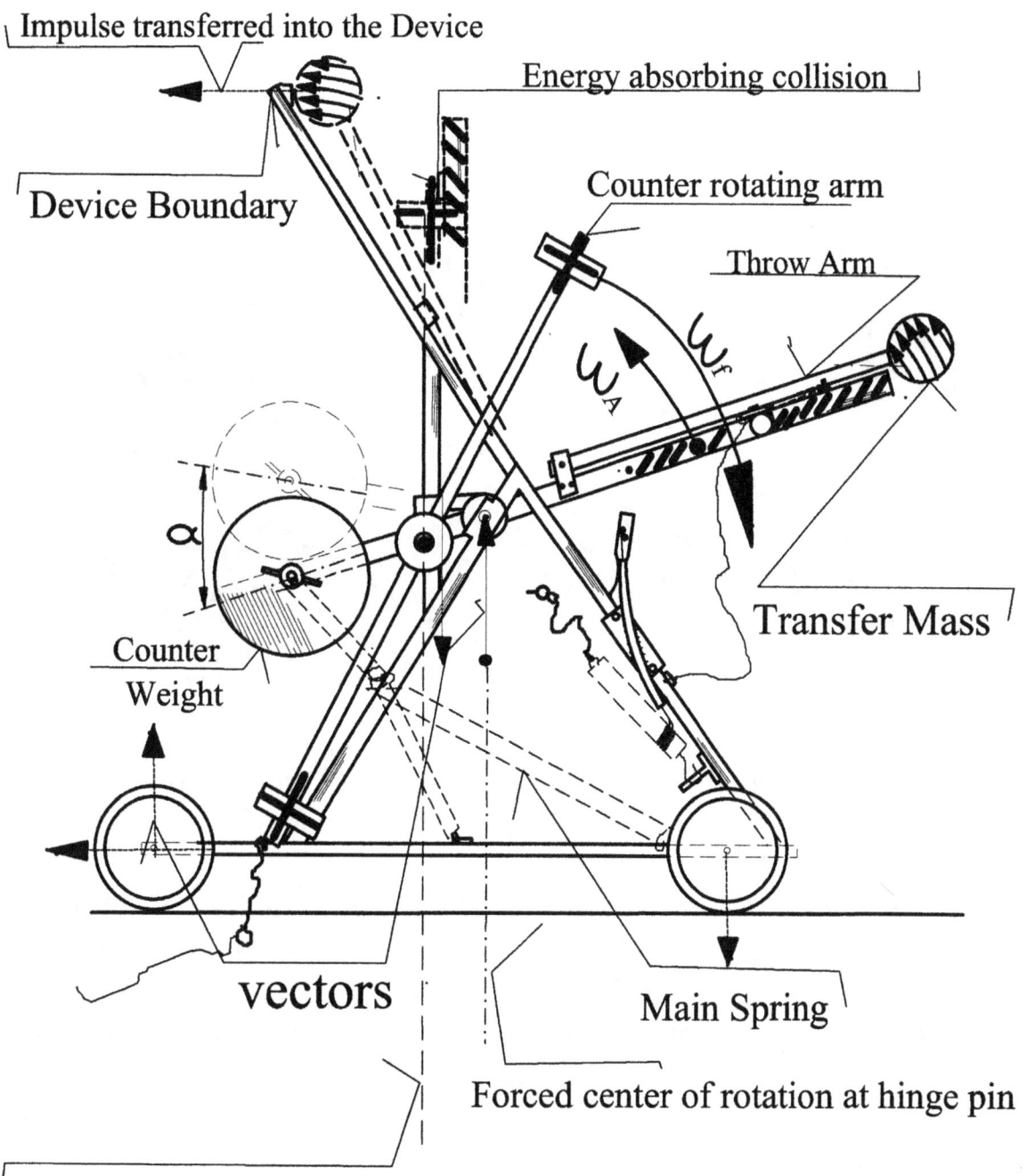

Impulse transferred into the Device

Energy absorbing collision

Device Boundary

Counter rotating arm

Throw Arm

ω_A

ω_f

Transfer Mass

Counter Weight

α

vectors

Main Spring

Forced center of rotation at hinge pin

Stationary natural center of rotation at CM

US Patent: 8491310

Next is a picture of the pendulum test of the Education Device displaying a self-contained impulse of more than 1 Newton-second. The impulse is lifting a 7 foot (2.13m) long, 8.3 Kg total weight pendulum and displacing the pointer tip 4.5 cm. The energy flow applied to the total inertial mass of the pendulum, when viewed in relation to the Saturn Rocket propulsion is 4.3 Kw energy flow. The picture is a snapshot of the www.mindbites.com/series1278/lesson7/ video presentations.

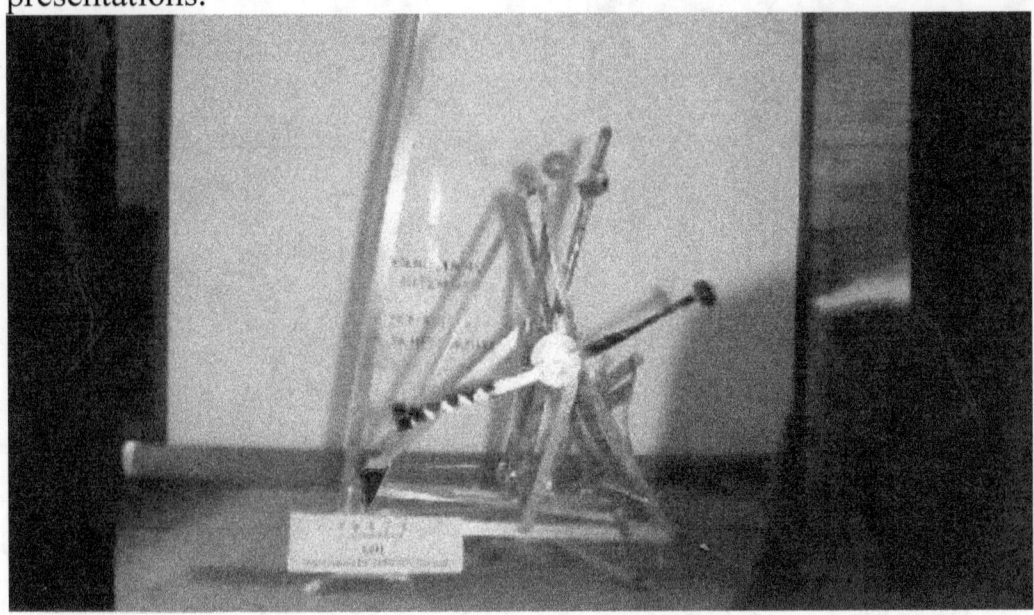

In view of the modified trebuchet pendulum test it should be possible to remove the objection based on the momentum and kinetic energy conservation principles in relation to Inertial Propulsion from our science books. From this example we are able to postulate again:

Within machines we are able to force impulse and momentum to following energy transactions. We are able to apply the conservation principles in terms of the isomorphic symmetry between impulse and energy. This is the reversibility of formula #6 and #8

"Why are we certifying the transfer of potential energy into isomorphic inertial mass motion momentum applying to formula #6 as satisfying the conservation principles; while in contrast, we assign invalidity to the isomorphic conservation of momentum into potential energy, which is the reversal of formula#6. This, we say, violates the conservation principles. This convention is violating the reversal of Physics principles and it prevents us from effectively engineering dynamic regenerative breaking systems and IP systems?". Because the center of mass received Ke, it must be accounted for!

In View of the performance reality of the modified trebuchet self-contained impulse example we must include the gain of kinetic energy applying to the whole aggregate inertial center of mass of the isolated IP system. We must include the Ke gain of the whole aggregate inertial CM mass into the Hamiltonian energy tally count for every IP cycle; we must update the isomorphic relation of impulse to Kinetic energy presented on page 64 to include the NEW formula:

Formula # New: $E_{\text{internal storage depletion}} - Ke_{\text{gain, system}} - E_{\text{friction, internal}} = 0$

The vertical Impulse of the Pendulum Swing

When observing the operation of an eight seat playground swing-set which has one child sitting motionless on the first seat and a vigorously swinging child on the end seat; we will notice that the swing set frame is knee-jerking at the end of the vigorously swinging child absorbing significant impulses. This phenomenon is related to the presented Inertial Propulsion tethered inertial mass mechanisms and needs quantifiable investigation with the applicable formulas. We investigate this phenomenon using a pendulum clocks on a weight scale. Let us now place a working-ticking pendulum clock and a stopped-dead pendulum clock on a weight scale and examine the difference of thrust between a stopped pendulum and a swinging pendulum.

We know that the instant centrifugal force is: $F = m_{\text{,pendulum, bob}} V^2 / r_{\text{,pendulum,rod}}$
We also know that within the gravitational acceleration the velocity in respect to the dropped height is applying to the repacking of formula #1:

$$V^2_{\text{gain}} = 2gh$$

At the bottom of the pendulum swing we arrive at an instant peak force of:

$$F_{\text{peak,centrifugal}} = m2gh/r$$

The averaging factor for an impulse for every position is according to the Root mean square method depicted in the following picture is: $2/\pi$. The method is the geometry of a geometric cut through the circle in a flat plane, wherein the cut length is the time and the segment height is the peak force further explained in **page#109**. Then the average centrifugal force is: $F_{\text{average}} = m4gh/r\pi$
The time of the swing is the invariable ratio of the radius to the gravitational acceleration, a repacking of formula #1.6: $T = 2\pi(r/g)^{1/2}$

The impulse of the pendulum is a function of the average centrifugal force:

$$P = F_{\text{,average}} * \text{Time} = m8gh(r/g)^{1/2}/r$$

The impulse is in addition to the average gravitational weight.
This example helps to explain the self-contained impulse generated by the

presented Inertial Propulsion device using the tethered orbit of an inertial mass. The tension spring can be of a soft deformable material proving the ability of the **centrifugal force performing real work** wherein the upswing height of the pendulum reduces by the spring deformation work performed!

The vertical Impulse of a Pendulum

gravitational   weight   Impulse

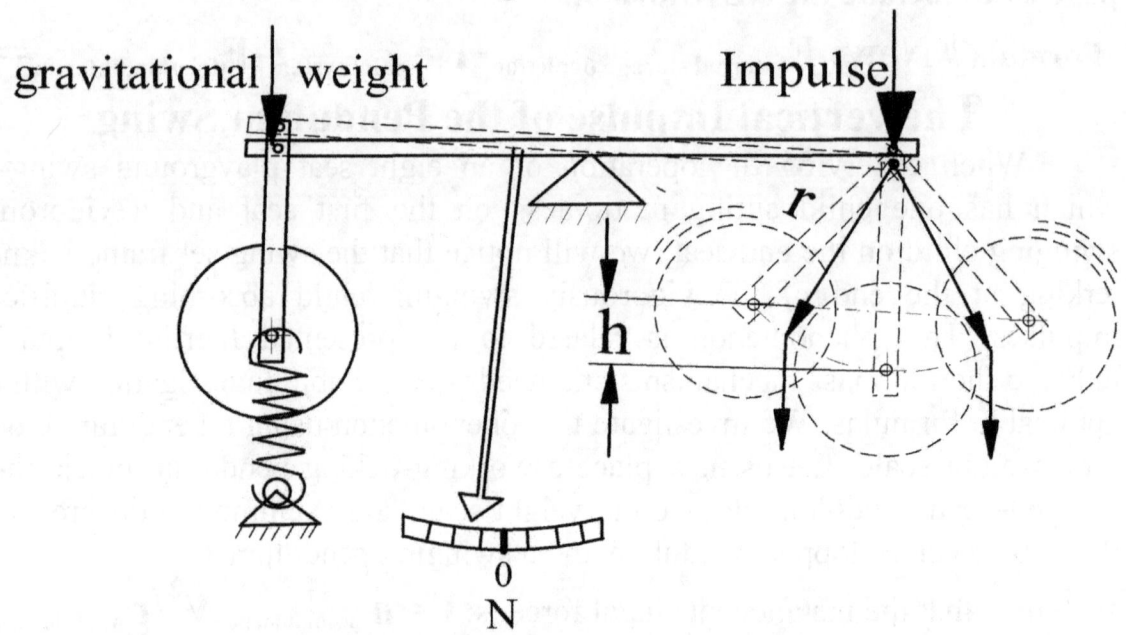

gravitational weight(N) = mass * acceleration

Impulse = centrifugal-force-average * time+weight-average

$$Centrifugal\text{-}force = mV^2/r$$
$$V^2 = 2gh$$

The Root Mean Square averaging:

$$\sqrt{}\,time = 2\pi\,\sqrt{r/g}$$

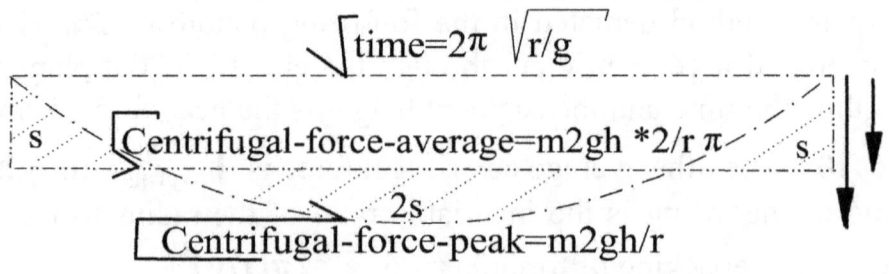

$$s \qquad \sqrt{}\,Centrifugal\text{-}force\text{-}average = m2gh*2/r\,\pi \qquad s$$

$$2s$$

$$Centrifugal\text{-}force\text{-}peak = m2gh/r$$

$$Impulse\text{-}centrifugal\text{-}average = m8gh/r\,\sqrt{r/g}$$

$$pendulum\text{-}weight\text{-}average = mg2/\pi$$

Pendulum, thrust=m4gh/rπ+mg2h/rπ

When we consider the pendulum swing height from h=0 to h=r; then the pendulum thrust is progressing from a balance to 1.9 times the weight!

Here is the pendulum in an oblique lean suspended on a long pendulum to achieve a horizontal self-contained impulse against its own CENTRE of MASS dependent on the ratio of the larger Mass **M** attached to the pendulum pivot in relation to the mass **m** of the pendulum bob in reference to the previous picture and in reference to Formula#8, example#5.

The the horizontal drive impulse
of the oblique pendulum dependency on E/e=m/M

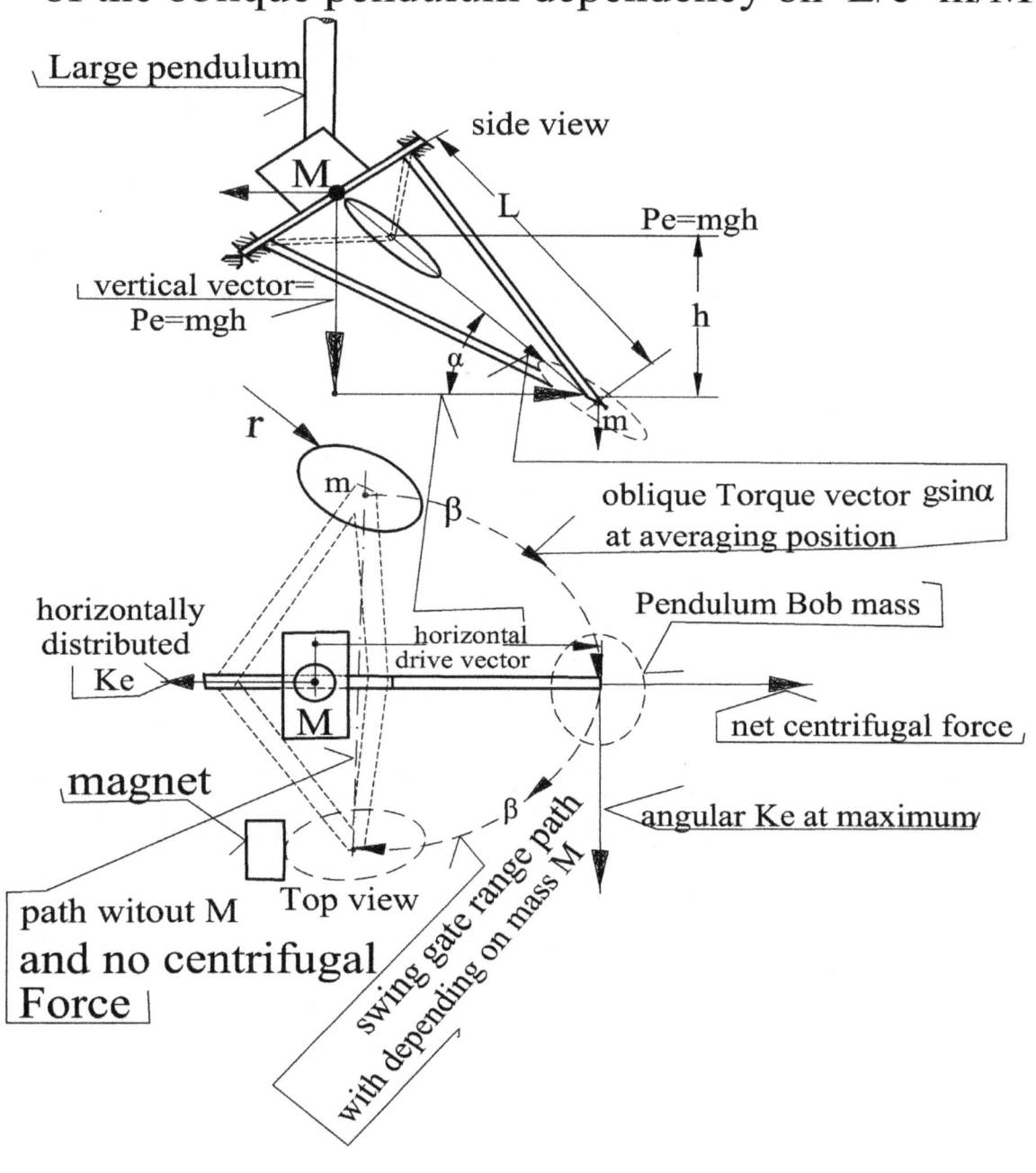

The first original root cause of the pendulum motion peak tangential velocity is the Potential Energy, which is the Huygens's **Pe=mgh**; from there we get the angular form of the pendulum motion caused by the **Pe** to **Ke** exchange, wherein the

$$Pe = mgh = Ke = I_{moment\ of\ Inertia}\omega^2/2.$$

The moment of inertia is $I=m(L^2+\frac{1}{2}r^2)$ because we have to account for the radius of the flat pendulum bob. The angular **Ke** is depending on the formula#8) **m/M=E/e** ratio and the total vertical **Pe=mgh** to develop the angular and the centrifugal mass motion of the pendulum moment of inertia **I**; accordingly, the larger the inertial mass **M** attached to the pendulum pivot in relation to the bob mass **m** the larger is the average centrifugal force up to an approximate maximum average:

$$F_{centrifugal,average,max}=mgh4/r\pi=2mV^2_{tangential,average,max}/r\pi.$$

The net centrifugal force is motivating the combined **M+m** center of mass (CM) in a self-contained inertial propulsion. For reality prove of the self-contained impulse against the Centre of the combined Mass; a video is provided on **"youtube ggutsche1 pc130005"** showing a single sweep of the oblique pendulum performing real work lifting the long suspension pendulum and causing a permanent swing of the long pendulum together with the oblique pendulum of **6.2cm** when attached to the magnet comprising a **3.7Kg** total pendulum bob mass after completing the single oblique pendulum sweep. This indicates that a combination of centrifugal force and the depleting of rotational kinetic energy of the pendulum bob **is** causing a gravity induced inertial propulsion; this is congruent with the skaters, trebuchet and the next presented continuous rotating IP, it is Newton's exception to his third law. Important to Note: The oblique pendulum presents the Physics case where we apply the centrifugal force mutual and reciprocally with an opposing straight line force applying to formula**#1**, wherein the centrifugal force performs the lifting of the total pendulum mass; this is similar to sample #7 were we see the distribution of the first root cause energy magnitude according to formula #8 with Newton's third law in the energy form! The understanding of physics teaches that the centrifugal force is a fictitious force unable to perform lifting; but obviously, in conjunction with the **third law in the energy form it is** performing **lifting,** because there is only one single Formula #8 centrifugal force magnitude doing the actual lifting and it is the distribution of energy by an energy force.

Is it possible to recharge the potential energy **Pe=mgh** each cycle without impinging an equal reciprocal impulse contrary to the self-contained impulse and obtain a repeating oblique swing and an repeating impulses?; yes, of course, an electromagnetic push solenoid can be used to lift the pendulum at each height amplitude back up to the original height potential and at the same time push the large main suspension pendulum nether forward nor backward (throw of an inertial mass within a vehicle), as presented in the next picture:

The the horizontal drive impulse
of the oblique pendulum with electrical solenoid driv

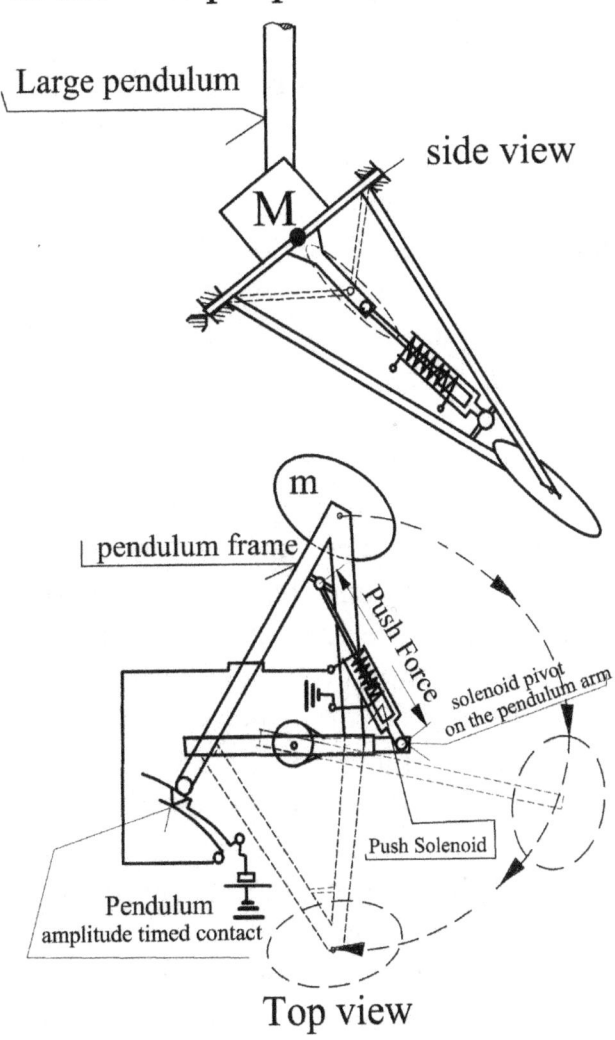

The Basic Requirements of Inertial Propulsion

The basic operational principle of Inertial Propulsion is based on Newton's time domain analysis of impulses, at the same time, the root cause of the impulses is based on Galileo's-Huygens-Leibniz-Lagrange displacement domain analysis. Newton's time domain analysis is requiring the generation of a unidirectional motivating self-contained energetic force impulse (Thrust) within a vehicle, in the direction of the intended motion of the vehicle. A self-contained impulse is self-contained if there are no force exertions against a fixed point external to the vehicle.

The Galileo-Huygens-Leibniz displacement domain analysis requires that the original root cause of the impulse is an internal self-contained source of energy quantity. The internal source of energy quantity is the work of an internal motor force over a distance. The force impulse must be regarded as the motivating agent of the isolated system of the vehicle and is the product of force and time interval applied to the whole aggregate center of mass of the vehicle. The internal product of force and time must be larger in direction of the intended motion of the vehicle to propel the vehicle forward. These principles indicate that not one single analysis procedure standing by itself can solve Inertial Propulsion problems. All principles are needed to quantify the internal impulse.

The presented Inertial propulsion drive is also employing a Huygens-Steiner Theorem type dynamic process using the combined effort of the two vector dimensions of the inertial reluctance contained in the mass motion of flywheels, the straight line displacement and the angular (rotational) reluctance to motion within a geometric plane. These principles were invented by Huygens in his "oscillaturium" paper. The dynamic process generates a timely sequential variable impulse mutually reciprocally exerted between the combined straight line and the rotational inertial mass reluctance of a flywheel and the aggregate sum of the Vehicle mass. The cyclic dynamic process further generates three timely, repetitive, identical, (base) initial mass motion potential energy conditions and one superior peak initial potential energy condition in a closed loop mutually reciprocal energy flow. This means that the timely sequential impulse has a superior magnitude in the direction of the intended motion of the vehicle and is applying Newton's first law: "The aggregate inertial mass of a Vehicle remains in motion until acted on by a subsequent superior opposing force impulse". These principles indicate that not one single analytical procedure, standing by itself, can solve Inertial Propulsion. All principles are needed to quantify the self-contained internal impulse. However, all four methods of analysis are important for IP depending on the physical environment the Inertial Propulsion vehicle is in. While a vehicle is within an intense gravitational field, the analysis must be in the time domain. Because the vehicle is not moving, the play of forces are only countering the gravitational force (hovering and the pendulum stall at a position) and all kinetic energy flow quanta is being recycled within the vehicle. Because of ZERO MOTION of the vehicle, except friction and efficiency losses of the moving Internal inertia elements, one can postulate that the generated force holding the vehicle in the hovering position is a net ZERO energy consumption; when discounting friction losses; this principle is applying to the pendulum test. When the vehicle is in a relatively low gravitational field the analysis must be in the displacement domain and in the time domain, because the vehicle is moving and is performing work against the force of gravity all at the same time. Thereby, the vehicle is displacing each quanta of kinetic energy per time frame (per operational cycle) and, therefore, the aggregate sum of the vehicles' masses is absorbing kinetic energy. This very important principle and its foundations are proven in the prove section of the publication. The exception to this simple rule is the consideration of the thrust timing that each cyclic dynamic process is delivering per vector dimension of inertial mass motion. This consideration has to be entered into the analysis. If the effective thrust timing is less than continuous, these time gaps will create a flow of energy between vertical (perpendicular opposed to the gravitational pull) potential energy and vertical kinetic energy of the vehicle, resulting in a vertical vibration. This vertical cyclic vertical vibration of the vehicle requires energy flow to sustain it. This should not be confused with energy consumption. The vertical vibration can be compared to the act of continuously kicking a ball up a steep hill. We will prove in this publication how this kicking the vehicle up a steep hill action and how the suspension from a pendulum affects inertial propulsion and the breakeven energy flow magnitude.

With these principles we therefore have proven that the presented device does not, <u>in any way</u>, generate free energy or sink energy into nothing; energy is always conserved within these devices as its own aggregate CM mass motion Ke; Publications assigning such physics incongruence to IP must be urgently updated!

The Inertial Propulsion Machine with Continuous Rotation

With the rotational dynamics of the skater, the trebuchet and the oblique pendulum sweep example proving a single self-contained impulse **is** possible, we now will proceed to a continuously cycling Inertial Propulsion process. The IP process derivation is accomplished with the Langrarian, Hamiltonian and displacement domain formulation of the original energy root causes of the inertial mass motions. The reality of the self-contained impulse is then proven with Newton's differential equation of the harmonic motion derived from the time domain analysis. Within the one single cycle skaters and modified trebuchet IP machines process we completely arrested the angular kinetic energy and the rotational momentum of the primary impact rotor (Flywheel #1) with a mutual and reciprocal energy absorbing collision based on formula #8 against a secondary rotor (Flywheel #2). In contrast, we now need a gentle mutual and reciprocal reduction of the rotational kinetic energy of the primary rotor, presented in Fig.1 as Flywheel 1 on page 107, by the secondary rotor (Flywheel 2) to maintain a continuous cycling (rotating) angular speed of the primary rotor; accordingly, the mutual and reciprocal absorbing energy flow **Essence** of the IP process presented for the skater and the trebuchet examples stays the same.

The primary working process of the presented continuous cycling Inertial Propulsion Device, presented on page 114 Fig.1, is Newton's straight line to rotational coupled (reflected) mass motion. When IP is implemented in its simplest form, a straight line movable crank shaft containing a rotational **Kinetic energy charged primary rotor** (flywheel 1) coupled to a straight line sinusoidal reciprocating mass motion by means of a straight line to rotational coupled Transmission Mechanism; this principle was already presented in example **#7**. We must consider that each ¼ cyclic primary rotor turn corresponds to one reciprocal motion direction and each ¼ flywheel 1 turn, and has a different average angular speed magnitude based on formula#1, #1a, #2 as presented in the beginning of the displacement domain analysis. The rotational to straight line coupled transmission is a mechanism converting rotation kinetic energy into straight line displacement inertial mass motion following a cyclic fixed progression of transmission ratios. Newton called this rotation to straight line displacement coupled motion very descriptively "straight line reflections of rotating motions". For our example in Fig.1 on page 118, we use the Scotch Yoke which follows a true sinusoidal transmission ratio. The straight line to rotational coupled mass motion is of premier importance because the prime originating source of the straight line shot put motivating force is Newton's steady continuing centripetal rotational force vector generated from the rotational arc motion of the rotor-crank-pin applying formula #9):

$$Z = Force_{,instantaneous,centripetal} = mass * Velocity^2_{,tangential} / Radius_{,gyration}:$$

Ref. www.wikipedia.org/wiki/Centripetal_force

The centripetal Force is denoted in good Physics books with "Z" because it is a complex function. Within formula #9 the tangential $Velocity^2$ is converted to angular velocity by multiplying the Velocity^2 by radius^2. We can do this because $\omega r = V$, by squaring both sides we get $\omega^2 r^2 = V^2$. Then we get $V^2/r = r * \omega^2$; then Formula #9.1: $\text{Force,}_{centripetal} = m_{inertial} * r * \omega^2$

Formula 9.1 is the rotational solution to the differential equation:

$$f(a) = d^2 s / t^2$$

The most important aspect of the centripetal force is that the energy stored in this rotational system is the centripetal force times the rotational displacement distance; it makes the rotational system an energy storage and energy supply device. In congruence, Newton's pendulum derivation taught us that we are compelled to divide the rotational displacement into equal parts arriving at a displacement type analysis of V^2; because the circle cannot be divided into unequal parts for analysis of the centrifugal acceleration. The Importance is that the rotational displaced distance traversed has a time duration proportional to the average angular speed. We have a force and time derived from speed; when both are multiplied we have impulse. Accordingly, it is a proven fundamental Physics principle to designate the rotational to straight line displacement transmission energy to momentum converter applying to formula #6. This conversion process has already been thoroughly proven with the Conveyor, Skaters, Gyrobus and the Trebuchet experiments. Furthermore, within the centripetal force there is no mass motion speed gain (speed amplitude) to average speed relationship, as we have seen in the straight line displacement mass motion of the race horse in formula #1a; only the magnitude of the tangential velocity vector magnitude in relation to a constant radius is causing the centripetal acceleration. Accordingly, we can say: The average speed, within an arc motion, is imbedded in the vector velocity divided by the radius. The radius parameter is in-extricable, in-disentangle-able from such a rotational motion system by any analytical view or any formula manipulation, no matter whether we are looking for impulses or for energy storage applications. The averaging factor question pertaining to the mean value versus the Cariolis averaging, we investigated for uniform verses non-uniform motion in formula #1a, does not apply here. This is because the averaging factor cancels out for the mean value impulse calculations: The force uses the mean value of angular velocity and the time duration also uses the mean value of angular velocity in an inverse relation, then both cancel out! The total time is divided into four parts to arrive at the ¼ turn time duration:

$$\text{Time, ¼ turn} = 2\pi/(\tfrac{1}{2}(\omega_0 + \omega))4 \ ; \ \text{Force,}_{average} = \text{mass} * r \, (\tfrac{1}{2}(\omega_0 + \omega))^2 * 2/\pi$$

$$\text{Force} * \text{time} = P_{impulse}$$

$$P_{impulse} = m * r(\tfrac{1}{2}(\omega_0 + \omega)), \text{ for a ¼ turn}$$

The $2\pi/(\frac{1}{2}(\omega_0+\omega))4$ term canceled out with $(\frac{1}{2}(\omega_0+\omega))^2*2/\pi$

ω_0 = original angular speed ; ω = new angular speed

The underlying root causes of the centrifugal acceleration are the vector ratios of dimensions within the triangles applying to a tangential segment of a circle, wherein the segment squared length (the length of the cut through the circle rim 2CD) has an approximate uniform relation to the product of segment height (the segment distance to the circle rim AD) times the radius **r**, this is because the areas $(AB)^2=(2rBC+(BC)^2)$ are equal. Next picture is Huygens first original geometric **vector-kinematic** derivation of the centrifugal acceleration:

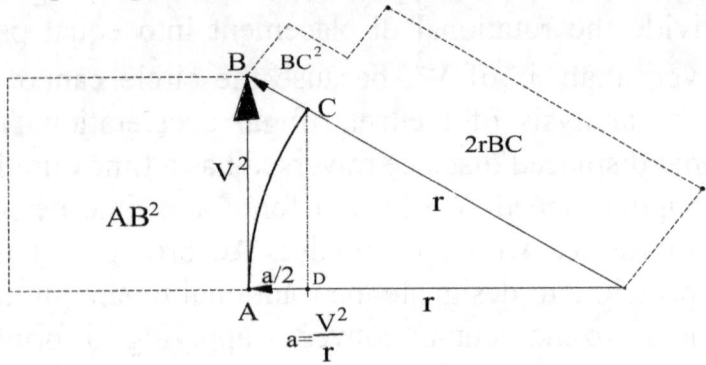

Wherein we assign congruence with the centrifugal parameters as follows: The length of the cut, **CD** = tangential velocity and

$$AD = \frac{1}{2}\text{acceleration, centrifugal.}$$

Accordingly, the centrifugal force is a displacement domain analysis related to formula #1 and #1c. The difference between centripetal and centrifugal is the direction of the arrow **BC**.

The kinetic energy stored into the angular inertial mass motion is independent of the force direction, centrifugal is the same as centripetal, and the **vector-kinematic** derivation from acceleration into the kinetic energy is the same as the straight line motion vector-kinematic of formula **#1a**:

$$E_{\text{energy,kinetic}} = ma_{\text{centrifugal}}\frac{1}{2}r$$

Then we must ask: "Where is the change from previous to new velocity and the displacement length within the centripetal acceleration?". For the centripetal acceleration the change from previous angular velocity to a new angular velocity and the displacement length is replaced by the radius magnitude in the denominator; this congruence is applying only to a constant angular speed.

The radius parameter is describing the magnitude of the redirection of the tangential motion vector for every change in angular displacement. A small radius has a large redirection per change in position.

The real difference of new and previous angular velocities applies only to a frequency modulated angular velocity used within the Inertial Propulsion Drive where we have a change in direction and also a change in angular velocity ω and a change in tangential velocity vector V within a ¼ turn. We can credit Newton for his fine performance to first recognize this very important relationship of circular motion and to recognize that it was a different concept than his straight line displacement force in the time domain. However, it was Huygens who promoted the importance of the displacement domain analysis and invented the Huygens Steiner theorem which is caused by to the centrifugal force. This principle is causing the rotating velocity vector magnitude to act both as velocity gain and also as velocity average, as seen in formula #1a, #4 and #7. Consequently, we must use the rotational vector magnitude Velocity^2, this is describing a quadratic functional system where energy and force are able to perform avalanche exchanges as we also experience in electrodynamics of collapsing coil magnetic fields and within the optic-dynamics of lasers.

From the steady continuing centripetal rotational force vector we extrapolate the straight line motion shot put force using a trigonometric vector projection (Newton's reflection) for the sinusoidal motion. Because the physics of the rotational to straight line displacement coupled motion has its fundamental principle in Huygens' centrifugal acceleration and Newton's centripetal acceleration, it also applies again, at the same time, to the Huygens-Steiner Theorem. This is caused by the variations of the Center of Mass within the total aggregate system and the in-variable center of rotation of the rotor, as we have seen in the trebuchet example, a quadratic functional relationship.

The centrifugal / centripetal force, a rotating vector force couple, has a steady continuing scalar presence, from a steady presence of angular velocity and is a quadratic magnitude of

Each angular velocity magnitude at each and every angular position incident of the rotational motion. This magnitude of acceleration per angular position (the angular value) theorem has been proven to be independent of any speed variations of the rotational motion over the motion distance; this means the centripetal/centrifugal force is always proportional to the angular velocity^2 **at the** local position.

Ref.: www.farside.ph.utexas/teaching/301/lectures/node89.html

The copy of the above proof of the University of Texas non-uniform angular motion is presented on the next pages, with their kind permission:

Non-uniform circular motion

Consider an object which executes *non-uniform* circular motion, as shown in Fig. 61. Suppose that the motion is confined to a 2-dimensional plane. We can specify the instantaneous position of the object in terms of its *polar coordinates* r and θ. Here, r is the radial distance of the object from the origin, whereas θ is the angular bearing of the object from the origin, measured with respect to some arbitrarily chosen direction. We imagine that both r and θ are changing in time. As an example of non-uniform circular motion, consider the motion of the Earth around the Sun. Suppose that the origin of our coordinate system corresponds to the position of the Sun. As the Earth rotates, its angular bearing θ, relative to the Sun, obviously changes in time. However, since the Earth's orbit is slightly *elliptical*, its radial distance r from the Sun also varies in time. Moreover, as the Earth moves closer to the Sun, its rate of rotation speeds up, and *vice versa*. Hence, the rate of change of θ with time is non-uniform.

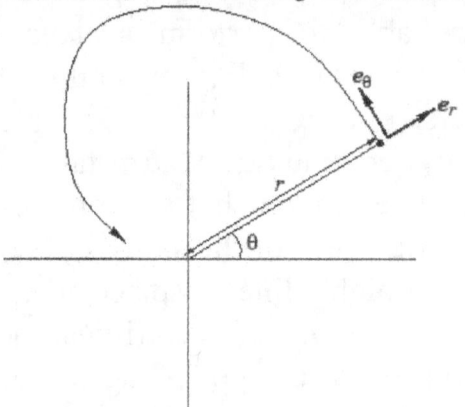

Figure 61: *Polar coordinates.*

Let us define two unit vectors, $\mathbf{e}_r$ and $\mathbf{e}_\theta$. Incidentally, a unit vector simply a vector whose length is unity. As shown in Fig. 61, the *radial* unit vector $\mathbf{e}_r$ always points from the origin to the instantaneous position of the object. Moreover, the *tangential* unit vector $\mathbf{e}_\theta$ is always *normal* to $\mathbf{e}_r$, in the direction of increasing θ. The position vector $\mathbf{r}$ of the object can be written

$$\mathbf{r} = r \, \mathbf{e}_r. \tag{270}$$

In other words, vector $\mathbf{r}$ points in the same direction as the radial unit vector $\mathbf{e}_r$, and is of length r. We can write the object's velocity in the form

$$\mathbf{v} = \dot{\mathbf{r}} = v_r\,\mathbf{e}_r + v_\theta\,\mathbf{e}_\theta, \tag{271}$$

whereas the acceleration is written

$$\mathbf{a} = \dot{\mathbf{v}} = a_r\,\mathbf{e}_r + a_\theta\,\mathbf{e}_\theta. \tag{272}$$

Here, v_r is termed the object's *radial velocity*, whilst v_θ is termed the *tangential velocity*. Likewise, a_r is the *radial acceleration*, and a_θ is the *tangential acceleration*. But, how do we express these quantities in terms of the object's polar coordinates r and θ? It turns out that this is a far from straightforward task. For instance, if we simply differentiate Eq. (270) with respect to time, we obtain

$$\mathbf{v} = \dot{r}\,\mathbf{e}_r + r\,\dot{\mathbf{e}}_r, \tag{273}$$

where $\dot{\mathbf{e}}_r$ is the time derivative of the radial unit vector--this quantity is non-zero because $\mathbf{e}_r$ *changes direction* as the object moves. Unfortunately, it is not entirely clear how to evaluate $\dot{\mathbf{e}}_r$. In the following, we outline a famous trick for calculating v_r, v_θ, *etc.* without ever having to evaluate the time derivatives of the unit vectors $\mathbf{e}_r$ and $\mathbf{e}_\theta$.

Consider a general *complex number*,

$$z = x + i\,y, \tag{274}$$

107

where x and y are real, and i is the square root of -1 (*i.e.*, $i^2 = -1$). Here, x is the real part of z, whereas y is the imaginary part. We can visualize z as a point in the so-called *complex plane*: *i.e.*, a 2-dimensional plane in which the real parts of complex numbers are plotted along one Cartesian axis, whereas the corresponding imaginary parts are plotted along the other axis. Thus, the coordinates of z in the complex plane are simply (x, y). See Fig. 62. In other words, we can use a complex number to represent a position vector in a 2-dimensional plane. Note that the length of the vector is equal to the *modulus* of the corresponding complex number. Incidentally, the modulus of $z = x + i y$ is defined

$$|z| = \sqrt{x^2 + y^2}.\tag{275}$$

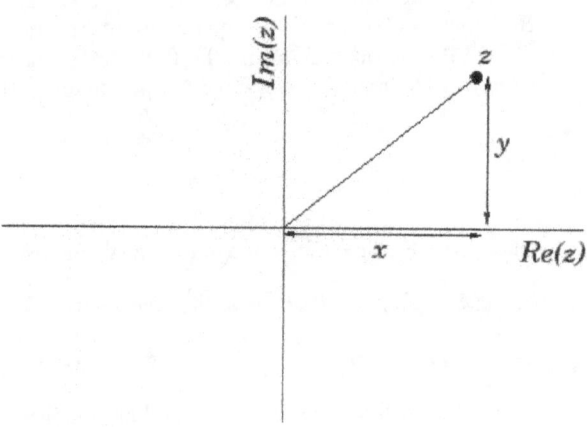

Figure 62: *Representation of a complex number in the complex plane.*

Consider the complex number $e^{i\theta}$, where θ is real. A famous result in complex analysis--known as *de Moivre's theorem*--allows us to split this number into its real and imaginary components:

$$e^{i\theta} = \cos\theta + i \sin\theta.\tag{276}$$

Now, as we have just discussed, we can think of $e^{i\theta}$ as representing a vector in the complex plane: the real and imaginary parts of $e^{i\theta}$ form the coordinates of the head of the vector, whereas the tail of the vector corresponds to the origin. What are the properties of this vector? Well, the length of the vector is given by

$$\left|e^{i\theta}\right| = \sqrt{\cos^2\theta + \sin^2\theta} = 1. \qquad (277)$$

In other words, $e^{i\theta}$ represents a *unit vector*. In fact, it is clear from Fig. 63 that $e^{i\theta}$ represents the radial unit vector e_r for an object whose angular polar coordinate (measured anti-clockwise from the real axis) is θ. Can we also find a complex representation of the corresponding tangential unit vector e_θ? Actually, we can. The complex number $i\,e^{i\theta}$ can be written

$$i\,e^{i\theta} = -\sin\theta + i\cos\theta. \qquad (278)$$

Here, we have just multiplied Eq. (276) by i, making use of the fact that $i^2 = -1$. This number again represents a unit vector, since

$$\left|i\,e^{i\theta}\right| = \sqrt{\sin^2\theta + \cos^2\theta} = 1. \qquad (279)$$

Moreover, as is clear from Fig. 63, this vector is normal to e_r, in the direction of increasing θ. In other words, $i\,e^{i\theta}$ represents the tangential unit vector e_θ.

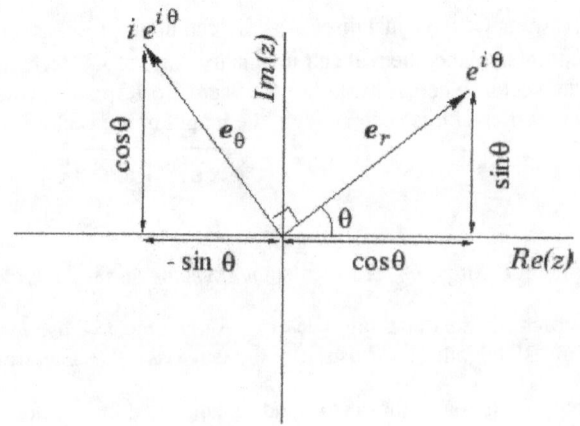

Figure 63: *Representation of the unit vectors* e_r *and* e_θ *in the complex plane.*

Consider an object executing non-uniform circular motion in the complex plane. By analogy with Eq. (270), we can represent the instantaneous position vector of this object via the complex number

$$z = r\,e^{i\theta}. \tag{280}$$

Here, $r(t)$ is the object's radial distance from the origin, whereas $\theta(t)$ is its angular bearing relative to the real axis. Note that, in the above formula, we are using $e^{i\theta}$ to represent the radial unit vector e_r. Now, if z represents the position vector of the object, then $\dot{z} = dz/dt$ must represent the object's velocity vector. Differentiating Eq. (280) with respect to time, using the standard rules of calculus, we obtain

$$\dot{z} = \dot{r}\,e^{i\theta} + r\,\dot{\theta}\,i\,e^{i\theta}. \tag{281}$$

Comparing with Eq. (271), recalling that $e^{i\theta}$ represents e_r and $i\,e^{i\theta}$ represents e_θ, we obtain

$$v_r = \dot{r}, \tag{282}$$

$$v_\theta \quad = \quad r\,\dot\theta = r\,\omega, \tag{283}$$

where $\omega = d\theta/dt$ is the object's instantaneous angular velocity. Thus, as desired, we have obtained expressions for the radial and tangential velocities of the object in terms of its polar coordinates, r and θ. We can go further. Let us differentiate $\dot z$ with respect to time, in order to obtain a complex number representing the object's vector acceleration. Again, using the standard rules of calculus, we obtain

$$\ddot z = \left(\ddot r - r\,\dot\theta^2\right)e^{i\theta} + \left(r\,\ddot\theta + 2\dot r\,\dot\theta\right)i\,e^{i\theta}. \tag{284}$$

Comparing with Eq. ([272](#)), recalling that $e^{i\theta}$ represents $\mathbf{e}_r$ and $i\,e^{i\theta}$ represents $\mathbf{e}_\theta$, we obtain

$$a_r \quad = \quad \ddot r - r\,\dot\theta^2 = \ddot r - r\,\omega^2, \tag{285}$$

$$a_\theta \quad = \quad r\,\ddot\theta + 2\dot r\,\dot\theta = r\,\dot\omega + 2\dot r\,\omega. \tag{286}$$

Thus, we now have expressions for the object's radial and tangential accelerations in terms of r and θ. The beauty of this derivation is that the complex analysis has automatically taken care of the fact that the unit vectors $\mathbf{e}_r$ and $\mathbf{e}_\theta$ change direction as the object moves.

Let us now consider the commonly occurring special case in which an object executes a circular orbit at *fixed radius*, but varying angular velocity. Since the radius is fixed, it follows that $\dot r = \ddot r = 0$. According to Eqs. ([282](#)) and ([283](#)), the radial velocity of the object is zero, and the tangential velocity takes the form

$$v_\theta = r\,\omega. \tag{287}$$

Note that the above equation is exactly the same as Eq. ([250](#))--the only difference is that we have now proved that this relation holds for non-uniform, as well as uniform, circular motion. According to Eq. ([285](#)), the radial acceleration is given by

$$a_r = -r\,\omega^2. \tag{288}$$

Consequently, in accordance with the **utexas** proof, the centripetal force is **directly proportional to the momentary kinetic energy content** of the rotating rotor mass rotational displacement and energy is accordingly the **original root cause** of the centrifugal- centripetal force couple. Importantly to note again is that the centripetal force is a force present for every position of a rotating structure congruent with formula#1. In contrast, Newton's force caused by straight line displacement acceleration is a mean value force over the time duration. Because the periodic cyclic motion of a reciprocating mass has both time duration and a centrifugal force magnitude for every position, then the periodic straight line displacement motion is also caused by an impulse. The impulse is the isomorphic symmetry of kinetic energy/work to impulse and also the force multiplied by the time duration. Here again most importantly, as we found with the Gyrobus, the mathematical representation of the rotational to straight line coupled motion cannot be described with impulse causing momentum, but with energy causing impulse considerations. This is because of the cyclic transmission of energy and the flow of stored kinetic energy in congruence with the Gyrobus. Then, the rotational to straight line coupled motion is related to the forces of the Hammer Throw Sport involving tethered mass motions and is NOT related to the shot put sport. "Which of these two sports gives the greater throw distance with the same body of energy?". The centripetal force is presented in text books, having for each and every individual straight line displacement position of the rotational coupled straight line motion, a force magnitude of:

$$\text{Force}_{\text{,instantaneous,}} N = \text{mass}_{,} * 4 * \pi^2 * S_{\text{,displacement,distance,straight,line}} / \text{time}^2_{\text{,cycle}}$$

When we convert the displacement to angular position we get:

$$\text{Force}_{\text{, instantaneous at position,}} N = \text{mass} * r * \omega^2 * \cos_{\text{,angle}}$$

Expressed in a differential equation:
$$f(a) = d^2s/dt^2 = a = -\omega^2 s$$

Ref: **Gieck formulas L7 Oscillations, L10 Scotch Yoke**
Reality check: www.farside.ph.utexas/teaching/301/lectures/node89.html

The maximum centripetal force is exerted when the straight line motion displacement position is equal to the rotor crank radius. This means, the instantaneous force is growing in a quadratic progression with the prevailing base cyclic frequency 1/time², cycle. The mean value force for each 1/4 cycle rotation of the crank is for a uniform (constant) rotation, the maximum instant centripetal force is multiplied by:

$$2/\pi=0.637$$

This averaging is a fundamental principle applying to any sinusoidal curve in inertial mass motion and in electrical curves:

Ref. www.electronics-tutorials.ws/accircuits/average-voltage.html

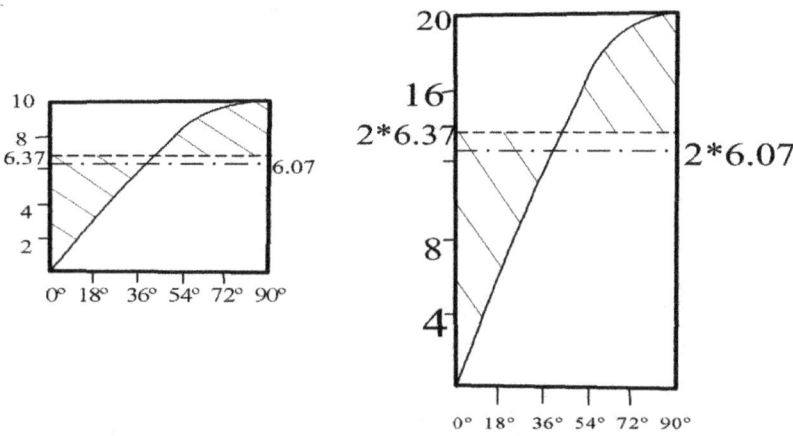

The shaded areas on the left and the right in each plot have equal magnitude. This is called the mean value theorem averaging factor; which is a **constant**. If we want to use the "Root Mean Square (RMS) averaging", we multiply the maximum instant force by:

$$\tfrac{1}{2}*2^{1/2}=0.707.$$

If we use the calculus based integral of sinusoidal surface area the modifier is:**1-π/8=0.6073,**this has the disadvantage of more complex derivation; because, then the **2/π**mean value parameter does not cancel out, a small disadvantage of a 0.03 factor inflated thrust magnitude in return for a much cleaner / simpler derivation.

It is important to remember that the modifier magnitude is a **constant** and independent of the calculated amplitude magnitude of the motion and the time rate of the motion as indicated with the graphic picture, because it modifies these magnitudes to arrive at the average per cycle; the calculation of the average force is the force delivered by each 1/4 turn of the rotor. Accordingly, the average force magnitude and the time duration is DIFFERENT for EVERY 1/4 turn **if** the average rotor angular velocity is different for the 1/4 turn while the amplitude is constantly repeating. This principle is telling us that there is a gradient self-contained mutual reciprocal energy distribution present for every 1/4 turn of the rotor. The rotational to straight line coupled motion follows a harmonious motion if the straight line motion velocity gain to the time ratio, which is:

$$V_{gain} = 2*\omega* radius/(time_{,duration\ per\ 1/4\ cycle}),$$

arrives exactly at each 1/4 turn of the rotor. The force averaging for non-uniform frequency modulated combined straight line displacement to rotational mass motion will be exhaustively proven using the average of the mean value theorem in the proof section following next. We will use the average angular velocity occurring at 1/8th of the 1/4 rotation and then apply the rise over run to this average angular velocity. "Is there an assumption in this method diminishing its general applicable validity?". The only assumption is that the frequency modulation is a steady uniform changing modulation for every increment of angular displacement. When we recheck the validity of the average of the mean value with formula #7, which we promoted to have the highest validity, we find that the force average and impulse returned is only slightly less using formula #7, but with absolute certainty is returning a self-contained impulse quantity within a frequency modulated inertial propulsion cycle. The average of the mean value has a very high validity status and ought to be generally accepted in school physics books but can only be found in bits and pieces therein.

The flow of kinetic energy for these 1/4 rotation examples is between the rotor-crank/cam and the straight line reciprocating mass, alternating reciprocally between the rotor-crank and the straight line displacing mass while the systems' total kinetic energy magnitude is conserved. Straight line displacement kinetic energy is flowing into rotational kinetic energy and vice versa, employing the two vector dimensions of mass motion. The energy flow for harmonious cyclic motion is basically identical to the straight line conveyor example; by using the mean value theorem we multiply the maximum force by the averaging factor $2/\pi$ thereby obtain the mean value of force for each ¼ turn. Force multiplied by velocity is power-flow; already proven with the conveyor:

$$\textbf{Power}_{average},\ \textbf{Kwatt} = \textbf{mass} * 8 * \pi*\textbf{Radius}^2_{crank}/\textbf{time}^3$$

The mechanical power-flow magnitude is a measurement of the required stencil strength of the mechanical components and motor size requirement. This is because the mass and the radius is **in**variable, the formula therefore means that: even a very instantaneous small localized variation of the cycle time duration will overwhelmingly, in a cubed progression, affect the internal power flow; the straight line motion impulse will be affected in a proportional relation in respect to the momentary angular speed magnitude per angular position. The impulse to power flow relation is a diminishing returns function in respect to the impulse applying to formula #6. Within discussions of IP principles, it can be found that this relationship is again and again misconstrued by an ingrained, incorrect, strange belief of proportional relationship of IP thrust to internal power-flow. Here we can present again: Power flow, in any system having delay reluctance is a function of the cycle frequency, which is a fundamental principle in Physics. Of course, this power to cycle frequency relationship is reversible. The systems cyclic frequency depends on the system's ability to deliver additional energy into the system. The energy demand is then a limiting factor of the IP thrust. The most important aspect of the mechanical translation of rotation to straight line mass motion is the inherent temporary kinetic energy storage and kinetic energy distribution cycle. The rotational kinetic energy of the crank rotor (flywheel 1) is distributed into the straight line reciprocating mass motion and the process reverses where the kinetic energy of the straight line reciprocating mass is accumulated into the crank rotor (flywheel 1). This statement, however, assumes that the rotor is stationary. When the rotor is having a straight line displacement and the rotor axis bears both the straight line displacement mass reluctance and the rotational mass reluctance, then the kinetic energy flow is distributed and re-combined between the rotational and the total straight line vector dimension of mass motion. The distribution ratio is then the reverse ratio of masses as previously presented in the separation of two unequal masses by the stored energy of a spring formula #8, #8.a, and it is a feedback principle. Because of this feedback principle, the impulse for every 1/4 turn is different if the initial potential energy condition of the angular speed is different at the start of the straight line displacement. Accordingly, we are at liberty to use formula #6 to convert the potential energy / work performed into an impulse magnitude, because this formula always returns the real usable portion of impulse. "Did Newton know this principle already?" "Was this the reason he stood clear within his principia of the rotational to straight line coupled motion?". Most likely, because all we have to do is couple a straight line guided inertial mass onto a pendulum through a connecting rod and there we can observe this rotation to straight line coupled motion flow principle in the time domain, as well as in the displacement domain. This is proving again, the flow principle of impulse / momentum by Hermann and Schmid. Now we can postulate with certainty that within the rotational to straight line coupled motion of the inertial propulsion device the performed

Impulse is an isomorphic function of potential energy.

Furthermore, Impulse is a proportional function of angular velocity.

When we look at the similarity of the rotational to straight line displacement coupled motion to the mutual reciprocal separation of unequal mass, we must point at the difference between the separation by the mechanical power of a spring and the separation by the power of a rotational inertial mass motion. The progression of the forces is caused by the energy flow; the product of force applied over a distance of motion per time interval. Considering that energy flow is the force multiplied by velocity, this relationship is summarized, and distilled into the universal applicable formula for inertial propulsion. The collapsing of angular velocity "A" dropping down to angular velocity "B" within a 1/4 turn, then back up to an angular velocity "C" per the next 1/4 turn is performing the inertial propulsion in sync with the back and forth reflected coupled motion of the transfer motion mass.

Referencing again Fig.1 page 118; the inertial propulsion is then performed by the difference between two angular velocities for a ½ turn within one complete turn of the primary impulse rotor flywheel 1. These principles will be proven in the "Proof Section" following next.

The net energy flow magnitude within a cycling system is: Formula # 10

#10) $\mathbf{Energy, Work = mass_{,motion,body} * radius^2 * \frac{1}{2}(\frac{1}{2}(\omega_{,a}-\omega_{,c}))^2}$

Ref: **Gieck Engineering Formulas Kinematics, Rotational Motion L 6, l-27**

The effective Net Impulse exerted by the device is the Isomorphic symmetry of energy and impulse in the real effective portion of the complex Cartesian Grid.

#11) $\mathbf{Impulse_{,net} = mass_{,motion,body} * radius * \frac{1}{2}(\omega_{,a}-\omega_{,c})}$

These final formulas are in the complex Cartesian grid and are universally applicable to all IP systems, including the authors Single Impulse Education Device. This formula can be verified and compared with formula #7. It will be shown in the body of the publication that both formulas provide a fair accuracy for the calculation of the propulsion impulse.

Then we must ask: "Where is the Gyrobus transmission ratio within formula #11?" The transmission ratio is the **radius** of the orbital motion <u>performing</u> the straight line to rotational reflection!

The Proof of the continuous Inertial Propulsion System

To present and prove the mutual separation of unequal masses by the rotation of an inertial mass moment we must present the modified scotch yoke mechanism side view in the next Fig.1, Fig.1b and top view in Fig.9 on page# 161 to formulate a mechanical mechanism which is actually capable of efficient continuous self-contained impulses. In Fig.1, the Scotch Yoke inertial mass straight line motion body consisting of the heavy combined mass of flywheel #1, flywheel #2 and the motor-generator. The flywheel #1 is mounted onto the motor generator drive shaft and the flywheel #2 is mounted onto the motor generator housings; this arrangement allows a mutual and reciprocal angular motion exertion between flywheel #1 and #2 on a common logical axis line in congruency with the skater, trebuchet and example #5, formula#8. Flywheel #1 is also called an impact rotor because it is the root cause of the self-contained impulse.

116

The straight line inertial mass motion body is being accelerated from **TDC** (the straight line motionless point) with a starting motion "**a**" by the elevated rotational motion kinetic energy level $\omega_{,a}$ of flywheel #1; this starting motion will be proven to be congruent with

the partial simulated dynamic expulsion of an inertial mass into empty space by exerting against the center of mass within a vehicle

without a subsequent negative reaction; because, the elevated flywheel #1 rotational kinetic energy has been absorbed mutually and reciprocally between flywheel #1 and #2 applying to formula #8 by the time the motion body reaches its starting motion mature velocity amplitude. In contrast, the flywheel #1 and #2 receive the rotational elevated potential kinetic energy level $\omega_{,a}$ mutually and opposing by the power of the motor-generator during the preceding stopping motion from the straight line motion amplitude in congruence with the skater, trebuchet and example #5 and formula#8, an **equal and opposing** actions. Here, again, there are no reactions forces at the stopping motion because it is consistent with an **inertial mass throw within the confines of a vehicle with equal reaction to an action**. After the starting motion the motion body is again de-accelerated from the straight line velocity amplitude by a straight line stopping motion; the two straight line stopping motion of the scotch yoke mechanism are again Newton's equal and reciprocal opposing action as presented in the preceding single impulse machine examples, it is conserving the motion body straight line kinetic energy amplitude back into flywheel #1, a self-contained dynamic braking system wherein the accumulator (battery) is the flywheel #1.

The inertial mass moment of the flywheel mass is variable in congruence with the trebuchet example. The inertial mass moment is depending, not only on the radius and the mass, but also on the orientation of the crank. When the crank-tether is in the horizontal position, the mass moment is purely a flywheel #1 moment. When the crank is in the vertical position the moment of inertia is a tethered (pendulum) inertial mass moment following the Huygens-Steiner theorem, formula #12:

$$M_{mass,total} = Flywheel_{,2,mass} + Motor,Generator_{,mass}$$

$$\#12) \qquad I_{,total} = I_{,flywheel,1} + r^2 M_{mass,total}$$

The kinetic energy of flywheel 1 at the start of the straight line drive phase starting motion was presented already in example #7 and is:

$$\mathbf{ke}_{flywheel,1,initial} = I_{total}\omega_a^2/2$$

The Lagrangian energy balance is according to the Gyrobus example on page 65 and is the working principle of the drive phase in Fig.1 and is formula

#1.b.1) $\qquad \mathbf{I}_{total}\tfrac{1}{2}\omega_{,a}^2 - \mathbf{e}_{pulse,motor} = \mathbf{M}_{mass,device,total}\mathbf{V}^2{}_{straight,line,gain}/2$

Accordingly, the angular speed amplitude $\mathbf{V}_{straight,line,amplitude} = r\omega_{,b}$ is variable; varying up and down to accommodate the stored kinetic energy in a varying magnitude of the inertial mass moment; this is also called a Frequency Modulated motion. A further increase of non-uniform angular speeds is accomplished with the pulsed torque exertions $\mathbf{e}_{pulse,motor}$ by the motor-generator exerting between the flywheel 1 and 2, independently of any other net exertions according to example#5 formula#8; the pulsed torque angular duration is in relation to the angular division distance of the motor-generator rotor. A pulsed torque, during the straight line stopping motion, is increasing the angular speed of flywheel 1 from ω_c to ω_a; there is no net exertion against the isolated system because of the conservation of energy in congruence with the trebuchet, and the mutual reciprocal regenerative dynamic braking. This is shown with the gray force couple in Fig.1 and Fig.1b, and proved with vectors in Fig.2C. The ¼ turn pulsed torque magnitude requirement is according to formula #1 force in the displacement domain rearranged into the angular form:

$$\boldsymbol{\tau}_{torque} = \mathbf{I}(\omega_a^2 - \omega_c^2)/\pi; \text{ wherein ¼ turn angular displacement}*2 = \boldsymbol{\pi}$$

The four starting and stopping motions of the total cycle are having two AC powered 1/4 cycles of the total. One 1/4 cycle is having one positive drive and one a negative drive. The remaining two cycles are idle rotations without any application of power, completing the cycle. This is indicated in Fig.1b and Fig.4 with drive cycles and idle cycles. In Fig.1 only the first stopping and starting motion of the complete cycle are shown. During the subsequent straight line starting motion a negative torque $\mathbf{e}_{pulse,motor}$ provided by the motor-generator reduces the angular speed by the mutual torque exertion between flywheel #1 and #2, causing large non-uniform angular speed, independent of any net exertions against the scotch yoke device frame. This is shown with the gray force couple against the device frame and the straight line motion of the flywheel - motor-generator mass applying the formula #8. This dashed------ force couple is a force couple for every position of the tether. Every change in position has time duration indicated with vertical time vectors for each position and therefore, an impulse magnitude. The force magnitude is a quadratic function of the straight line displacement section **S** of the total inertial mass of the flywheel-motor-generator straight line motion assembly and is a displacement analysis applying to Galileo's equal distance notched board, the equal angular division of the motor-generator rotor and formula #1. The average force, time duration and average impulse occur at 45° and are indicated with a vertical time vector length at every position. The guidance frame surfaces are forcing the direction of the vectors in congruence with sample #2. The horizontal propulsion forces for every position of the tether against the scotch yoke frame depicted in Fig.1 are the real component of the Complex Cartesian Grid while the straight line velocity vector is the time delayed lagging 90° imaginary component of the complex grid. This is because these forces are able to perform real work in the form of lifting an inertial mass and also frictional work which is generating heat. This has been proven with three examples: the skaters, the modified Trebuchet and the Gyrobus.

Here we arrive at the point where we diverge from Newton's world of kinematic understanding, Newton would have reject the notion that the square root out of $(-1)^{\frac{1}{2}}$ has any meaning. While today the idea of apparent power to real power charging differential by the electricity company is today readily accepted; therefore, the false notion of free energy device attached to Inertial Propulsion must be vigorously corrected, Inertial Propulsion is in fact the same as apparent power and real power differential; while the Newtonian impulse sum and resultant velocity gain sum is always true as long as enough apparent internal power-flow margin supports it. If the scotch Yoke is held stationary, the rotational kinetic energy accumulated in flywheel 1, during the stopping motion, is returned to the power supply. If the whole aggregate inertial mass of the scotch yoke frame moves-accelerates, then the difference between the positive torque and the negative torque multiplied by the rotational displacements is motivating the device because the difference in torque is causing a difference in directional impulse magnitude multiplied by the average angular speed.

Referencing now Fig.1, Fig.2, Fig.2b, page #118-131 the straight line motion instant force (Fig.2, the four dashed ------ lines) causing the starting motions are plotted for four different angular speeds starting at ω=100, 110, 120, 130 and 140 1/sec (the four solid _____ sloped lines) causing the starting motions in Fig.1. The 1/4 turn total time durations are the vertical -..-..-.. lines. The angular de-acceleration is congruent with a uniform angular de-acceleration presented in formula #1.6. A uniform angular de-acceleration produces a non-uniform straight line starting motion in plotted in Fig.2b. The force plot for each angular velocity position is derived by calculating the force at this particular position taking the changing time duration per position and the angular velocity magnitude into **simultaneous consideration**, thus the Fig.2 force plot is of high accuracy quality in congruence with the calculus accuracy requirement presented with formula #1a and in congruence with Newton's derivation of the accelerated pendulum on page #49, 50.

Fig.2b is the straight line motion velocity plot which is out of phase by 90° from the real force plot and has a real tangential line and a phase shift tangent line. Furthermore, Fig.2b indicates that it is important to analyse the physical mechanics in detail to determine the real portion and the complex imaginary portion within the inertial mass motions. The assignment of imaginary numbers to the vectors is not necessarily providing an accurate picture of the mechanical operation. It is then proven that the mean value of real Force in the real domain is always at the physical position of 45°, at this point having a force RMS averaging factor of $\frac{1}{2}2^{1/2}$ and a force mean value factor of $2/\pi$. This is indicated with the horizontal -.-.-.-. lines. This inter-correlation has already been proven to be independent of the angular speed non-uniformity progressions; whether $\omega_{,a}=\omega_{,b}$ or $\omega_{,a}>\omega_{,b}$ makes no difference, the final formula remains the same; therefore we observe that:

The area circumscribed by points ABE is the impulse magnitude and is greater than area of points ACD and we also observe that ABE=1.2ACD. This area relationship incontestably proves that the straight line Impulse for the ¼ (flywheel 1) turn performing the first straight line starting motion a of the cycle is equal to formula #13:

$$\text{Formula \#13:} \quad \mathbf{p_{,impulse}=m*r* \tfrac{1}{2}(\omega_{,a}+\omega_{,b})}$$

$\omega_{,a}>\omega_{,b}$ is plotted in four increments of 10 rad/$_{second}$, the largest difference between $\omega_{,a}$ and $\omega_{,b}$ is 40 rad/$_{second}$. The Net Impulse over the 360° total cycle displacement is accordingly proportional to the difference of the opposing ¼ turn impulses

$$\mathbf{p_{,net,Newton,second}=p_{,impulse,forward}-p_{impulse,backward}}$$

of flywheel **#1**; for the Fig.1, Fig.2 example the maximum possible net impulse is:

$$\mathbf{P_{max}=mr1/2(140-100)=20 \text{ Newton second}}$$

It also proves incontestably that the impulse is proportional to the average angular speed for every 1/4 turn of a spin if the spin is non-uniform; is variable at a constant rate applying to formula #1.6. This appears as a fundamental concept in nature which is congruent with photo-voltaic dynamics, thermos-dynamics and electro dynamics, therefore we have proven that:

Within the isolated system of machines impulses are caused by the first original root cause of energy flow. Or:
Inertial Propulsion is the internal mechanical energy translation into impulses from within.

Furthermore, because the motion body mass is unity 1Kg, the Fig. 1 plot is in reality a kinematic proof in congruence with the displacement domain kinematics! Additional importance: formula #13 is only considering impulse in relation to the angular momentum differential in a static inertial propulsion stall condition when the inertial propulsion machine is held in a **force stall condition**; if the system is transferring kinetic energy into the center of mass of the isolated system then we have to use the energy form, wherein the angular speed $\omega_{,c}$ is the T_{top} D_{dead} C_{center} of the opposing ½ cycle of Fig.1, the angular speed $\omega_{,a}$ is larger than $\omega_{,c}$:

#10)   Energy,Work=mass$_{,motion,body}$ * radius^2 * ½(½($\omega_{,a}$-$\omega_{,c}$))^2

Fig. 2c proves the independent relationship of the motor-generator modulation torque exerted mutual reciprocally and singularly between flywheel #1 and flywheel #2 according to Formula**#8** in Example **#5** the Third Law in Energy form. The flywheel 1 rotational frequency modulation must be regarded as a kinetic energy accumulation phase.

On the next page Fig.1 is depicted:

Fig.1
Modified Scotch Yoke

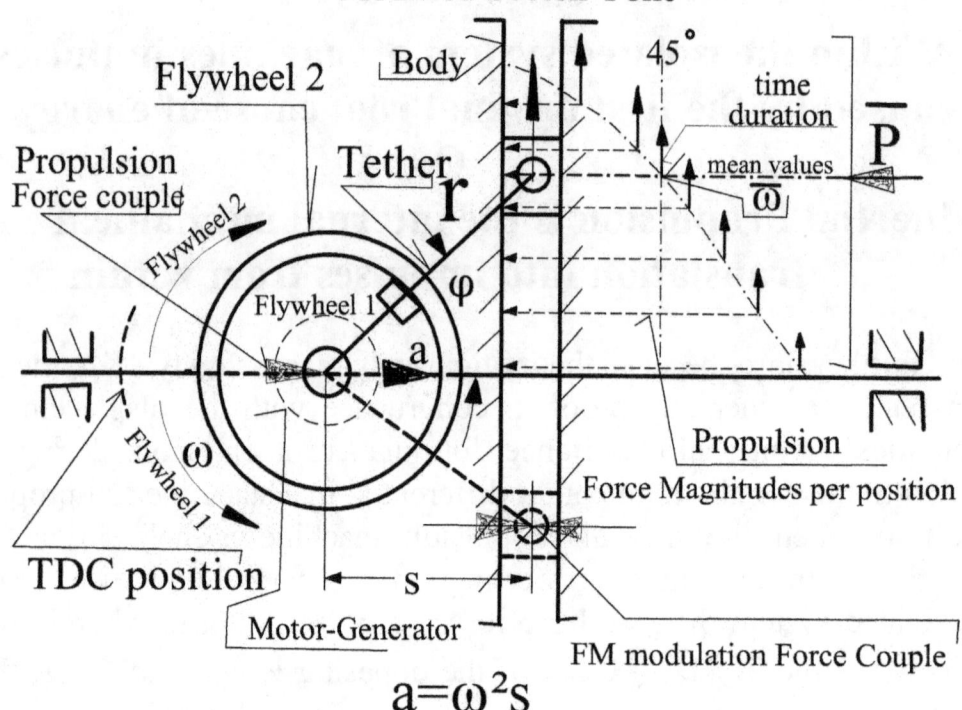

$$a = \omega^2 s$$

$$s = r * \cos\varphi; \text{ mean value } a \text{ is } \varphi = 45°$$

Flywheel 1 is mounted onto the motor generator shaft

The Tether (crank) is mounted onto Flywheel 1

The Motor-Generator housing is firmly embedded into Flywheel 2

The solid Tether (crank) is shown in a starting motion progessing from TDC to V amplitude

The dashed Tether (crank) is shown in a stopping motion from V amplitude to TDC

Two alternating FM modulating PWpulses are applied to the motor Generator

during the shown consecutive stopping and starting motions causing the FM modulation

Fig.2
1/4 turn impulse plot

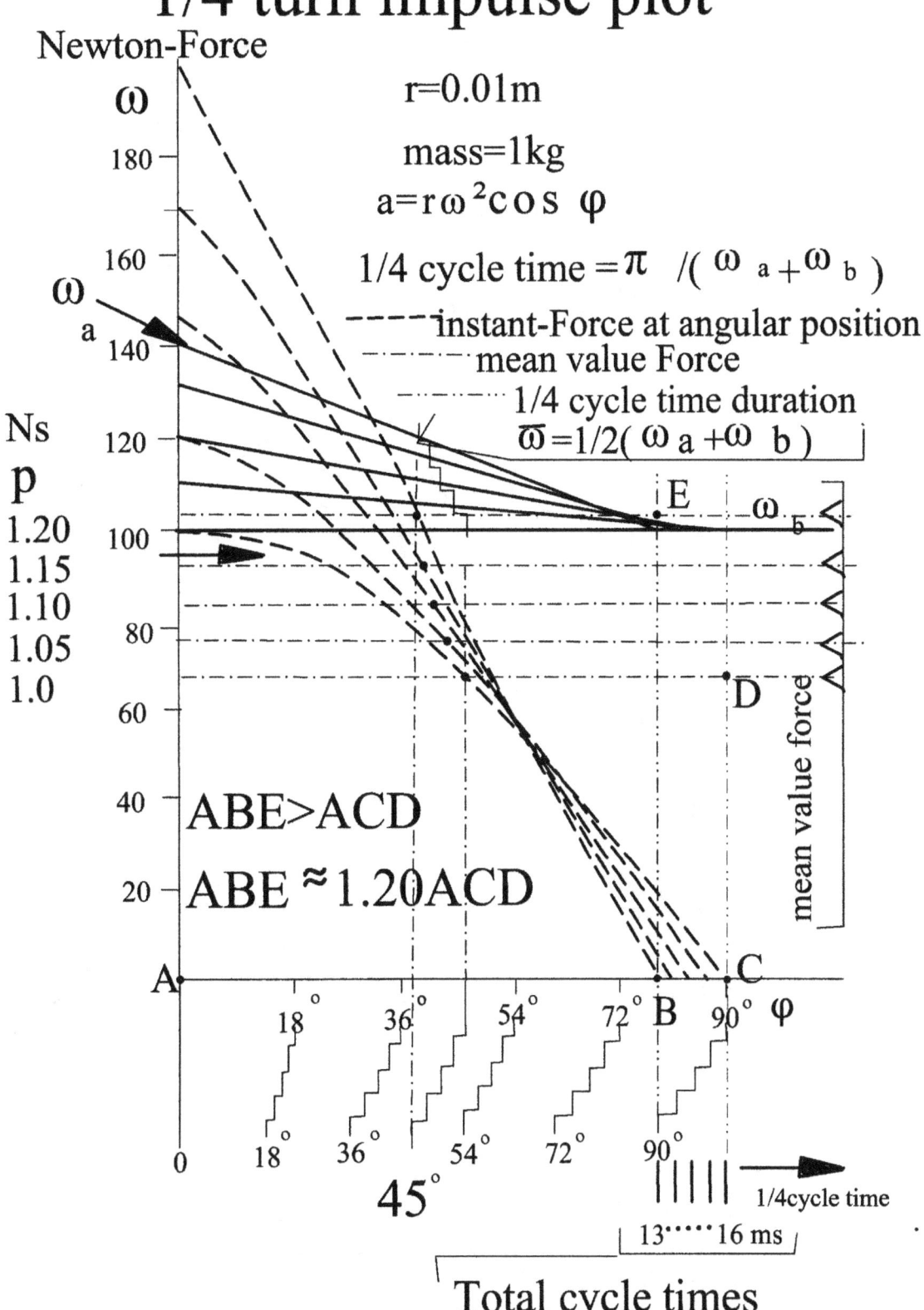

$r=0.01m$

$mass=1kg$

$a=r\omega^2\cos\varphi$

1/4 cycle time $=\pi/(\omega_a+\omega_b)$

- - - - instant-Force at angular position

- · - · mean value Force

· · · · 1/4 cycle time duration

$\overline{\omega}=1/2(\omega_a+\omega_b)$

ABE>ACD

ABE ≈ 1.20ACD

mean value force

$45°$

$13\cdots 16$ ms

1/4 cycle time

Total cycle times

Fig.3
Impulse Gradient per Angular Speed Gradient

Mass, linear, displacement =1Kg
Displacement, cam-motion, 1/4turn = 10 mm
Motion Impulse = Force * Time

Force, average, 1/4turn = mass * angular, speed^2 * displacement, linear
Impulse = mass * angular, speed * displacement, linear

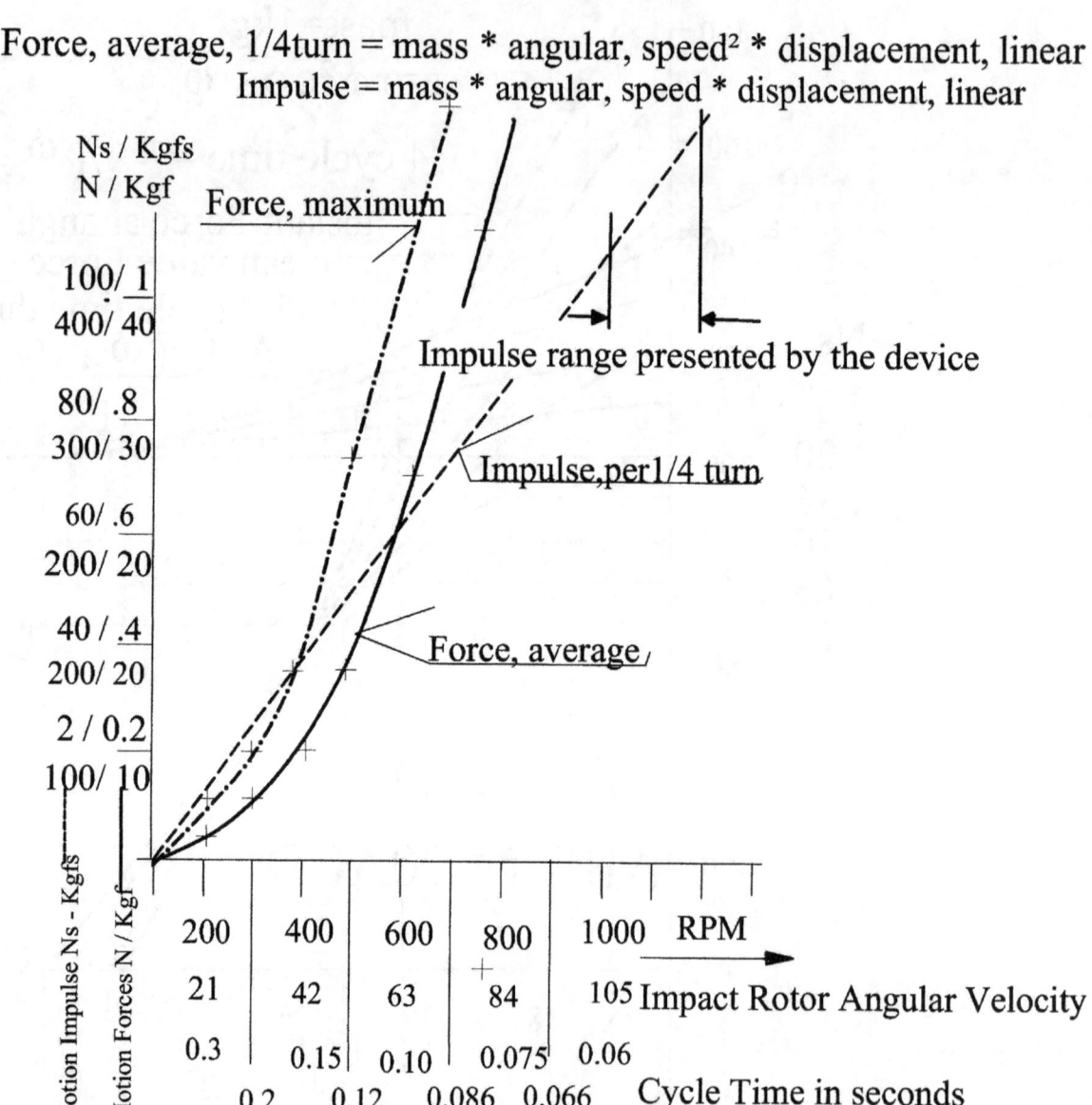

The above Fig.3 graph illustrates the quadratic functional character of straight line displacement forces at play, operating the combined effort inertia drive. This quadratic functional characteristic illustrates the challenge in scaling

124

the thrust up to very high propulsion forces. The forces are calculated for a total straight line flywheel assembly stroke of 20 mm and therefore, a radius distance of 10 mm for each acceleration and de-acceleration.

Furthermore Fig.3 illustrates the scalability of the inertial propulsion and the upper limit of this scalability. The straight line forces have quadratic functional character, which limits the possible maximum obtainable propulsion forces due to the limitation of the stencil strength of the construction material. The impulse is force multiplied by the time duration for every displacement section. The impulse is rising proportionally with the magnitude of the angular rotor velocity.

This graph illustrates the very important fact that the performance of the combined effort of inertial propulsion depends largely on the level magnitude of the angular impact rotor velocity, and is **only** limited to the maximum angular rotor velocity possible which is the stencil strength of the devices' motion parts. Ref. **Kurt Gieck Formulas L10, P10; www.epi-eng.com**.

Then it can be postulated with certainty:

The larger the average angular rotor velocity and the larger rotor angular rotational frequency modulation, the larger is the propulsion thrust in a diminishing returns progression. This means: An angular velocity in the thousands range is required to overcome the Earth's gravitational field.

Next three pages show a derivation of the Scotch Yoke inertial mass motion forces, impulses and kinetic energy from an alternate source. The pages are a copy from the book by Prof. G. Hausner of Caltech titled "AVANCED METHODS IN DYNAMICS. He calls the inertial mass motion "Forced Inertial Mass Motion". His derivations show the congruence with the Gieck formula derivation used for the preceding proof.

9.12 Forced Oscillations. We shall now suppose that the statically coupled conservative system which has been considered above is acted upon by a system of periodic exciting forces $F_i \sin \omega t$ so that the basic equations of motion become:

$$\frac{d}{dt}\left(\frac{\partial T}{\partial \dot{q}_i}\right) + \frac{\partial V}{\partial q_i} = F_i \sin \omega t \qquad (9.25)$$

Following the same procedure that was successful for the single degree of freedom system of Chapter 5, we shall assume that the steady state solution will be harmonic and of the same frequency as the exciting force. In what follows we shall consider steady state motion only. We suppose that there is a small amount of damping present which after a time will eliminate the transient terms, but which is not large enough to change appreciably the steady state forced amplitudes.

It cannot be expected that the solution will involve only a single mode of vibration, since all of the modes may be excited simultaneously. We accordingly write a trial solution in the form:

$$q_i = \sum_{r=1}^{n} c_i^{(r)} \sin \omega t \qquad (9.26)$$

where the coefficients $c_i^{(r)}$ are the steady state amplitudes of the forced oscillations of the various modes. We do not need to include a phase angle in this expression, since for the undamped system the phase shift will be either zero or 180°.

Substituting the general expressions for kinetic and potential energies into Equation (9.25) we obtain:

$$m_i \ddot{q}_i + k_{i1} q_1 + k_{i2} q_2 + \cdots + k_{in} q_n = F_i \sin \omega t$$

Putting Equation (9.26) into this equation, and cancelling the common $\sin \omega t$ term leads to:

$$-m_i \omega^2 \sum_{r=1}^{n} c_i^{(r)} + k_{i1} \sum_{r=1}^{n} c_1^{(r)} + k_{i2} \sum_{r=1}^{n} c_2^{(r)} + \cdots = F_i$$

or:

$$-m_i \omega^2 \sum_{r=1}^{n} c_i^{(r)} + \sum_{j=1}^{n} k_{ij} \sum_{r=1}^{n} c_j^{(r)} = F_i \qquad (9.27)$$

The next step will consist of expressing the coefficients $c_i{}^{(r)}$ in terms of the mode shape numbers $A_i{}^{(r)}$ that have been defined above in connection with the free oscillation problem. We define a new coefficient, $a^{(r)}$ as:

$$c_i{}^{(r)} = a^{(r)} A_i{}^{(r)} \qquad (9.28)$$

The physical significance of Equation (9.28) is that we suppose each mode excited by the external forces to have the same shape as the free oscillation mode, with an amplitude $a^{(r)}$ that remains to be determined.

Substituting Equation (9.28) into Equation (9.27), we find that the second term becomes:

$$\sum_{j=1}^{n} k_{ij} \sum_{r=1}^{n} c_j{}^{(r)} = \sum_{j=1}^{n} k_{ij} \sum_{r=1}^{n} a^{(r)} A_j{}^{(r)}$$

$$= \sum_{j=1}^{n} k_{ij}(a^{(1)} A_j{}^{(1)} + a^{(2)} A_j{}^{(2)} + \cdots)$$

$$= a^{(1)} \sum_{j=1}^{n} k_{ij} A_j{}^{(1)} + a^{(2)} \sum_{j=1}^{n} k_{ij} A_j{}^{(2)} + \cdots$$

$$= \sum_{r=1}^{n} a^{(r)} \sum_{j=1}^{n} k_{ij} A_j{}^{(r)}$$

We now refer back to Equation (9.23), which states that:

$$\sum_{j=1}^{n} k_{ij} A_j{}^{(r)} = m_i \omega_r{}^2 A_i{}^{(r)}$$

Introducing this into the last expression above, we can write Equation (9.27) as:

$$- m_i \omega^2 \sum_{r=1}^{n} a^{(r)} A_i{}^{(r)} + \sum_{r=1}^{n} a^{(r)} m_i \omega_r{}^2 A_i{}^{(r)} = F_i$$

$$\sum_{r=1}^{n} a^{(r)} m_i A_i{}^{(r)} (\omega_r{}^2 - \omega^2) = F_i \qquad (9.29)$$

Note that for the single degree of freedom case Equation (9.29) reduces to the form:

$$am(\omega_r{}^2 - \omega^2) = F$$

$$a = \frac{F/m}{\omega_r{}^2 - \omega^2} = \frac{F/m\omega_r{}^2}{1 - \left(\dfrac{\omega}{\omega_r}\right)^2} = \frac{F/k}{1 - \left(\dfrac{\omega}{\omega_r}\right)^2}$$

which checks the conclusions reached in Chapter 5 for the steady state amplitude of the single degree of freedom system with no damping.

For the multiple degree of freedom system this calculation is not so simple, because the amplitude coefficients $a^{(r)}$ occur in Equation (9.29) as the coefficients in a series. These constants can be determined, however, by expanding the exciting forces F_i into a series of the same normal functions $A_i{}^{(r)}$ that appear with the amplitude coefficients. We will thus write the exciting force as:

$$F_j = \sum_{r=1}^{n} f^{(r)} m_j A_j{}^{(r)} \tag{9.30}$$

where we use j instead of i to indicate that such an expansion could be applied in general to any one of the forces. This process of expanding the function F_j into a series is similar in principle to the expansion of a function in a Fourier series, but instead of sines and cosines we use the functions $A_j{}^{(r)}$ because of their appropriateness to the particular problem involved. The method of determining the unknown coefficients is analogous to that used for the Fourier coefficients.

Writing several terms of Equations (9.30) gives:

$$F_j = f^{(1)} m_j A_j{}^{(1)} + f^{(2)} m_j A_j{}^{(2)} + \cdots$$

Multiply both sides of this equation by $A_j{}^{(r)}$, and thus obtain:

$$F_j A_j{}^{(r)} = f^{(1)} m_j A_j{}^{(1)} A_j{}^{(r)} + f^{(2)} m_j A_j{}^{(2)} A_j{}^{(r)} + \cdots$$

Summing over both sides of this set of n equations gives:

$$\sum_{j=1}^{n} F_j A_j{}^{(r)} = f^{(r)} \sum_{j=1}^{n} m_j [A_j{}^{(r)}]^2 + \sum_{i=1}^{n-1} \sum_{j=1}^{n} f^{(s)} m_j A_j{}^{(r)} A_j{}^{(s)}$$

To positively illustrate the exceedingly complex IP drive principle of frequency modulated cyclic mass motion having a complex cyclic logic, a typical cyclic mass motion contour graph in 1/5 increments (sections) of the total cycle time-rotation is presented. Each 90° rotor rotation is having alternatingly 1/5 and 3/10 of the total cycle time duration. This timing represents the point of optimum cycle time fraction distribution; higher differential rate of distribution has slight diminishing returns of the impulse magnitude.

The areas under the displacement speed curves represent the straight line displacement length. This is congruent with Newton's time domain logic presented for formula #2. The **dv/dt** snapshot principle is used to plot the instant Force per each time position.

We observe in Fig.4 how the instant force per time duration and the velocity amplitude in respect to time is developed as an area under the velocity curve to indicate the complex velocity slopes of this mechanism.

Here is where we observe again how the instant force, the instant time duration and the total impulse and momentum area is developed visually proving a higher impulse area in the forward direction in comparison to the backward (reaction) direction! Therefore, it is again formally proven that the presented continuous rotating two opposing flywheel modified scotch yoke inertial propulsion device is bypassing the equal reaction to an impulse action theorem!

The motor-generator alternating drive pulses timing and amplitudes consisting of the positive drive pulses and the dynamic breaking drive pulses timing (the return of energy to the source) is also included in Fig.4, wherein the difference of the positive drive pulse and the negative dynamic breaking drive pulse is the kinetic energy induced into the centre of mass of the vehicle.

These principles are observable in the next Fig.4:

Fig.4
The Frequency Modulated Sinusoidal Mass Motion

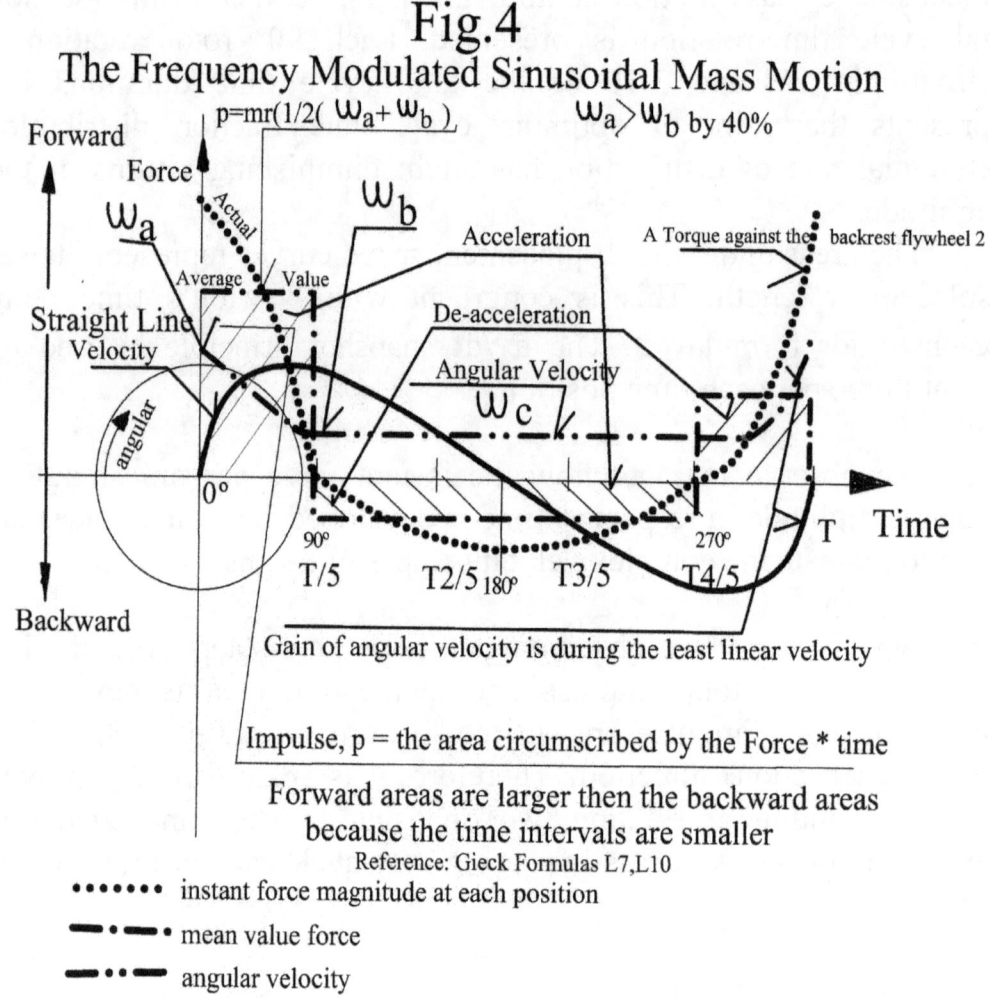

$$p = mr(1/2(\omega_a + \omega_b))$$ $\omega_a > \omega_b$ by 40%

ω_a

ω_b

Forward Force

Actual

Average Value

Acceleration

A Torque against the backrest flywheel 2

De-acceleration

Straight Line Velocity

Angular Velocity

ω_c

angular

0°

Time

Backward

90°   T/5   T2/5 180°   T3/5   270°   T4/5   T

Gain of angular velocity is during the least linear velocity

Impulse, p = the area circumscribed by the Force * time

Forward areas are larger then the backward areas
because the time intervals are smaller

Reference: Gieck Formulas L7,L10

•••••• instant force magnitude at each position

—·—· mean value force

—··— angular velocity

———— straight line velocity

The view of the continuing cycle

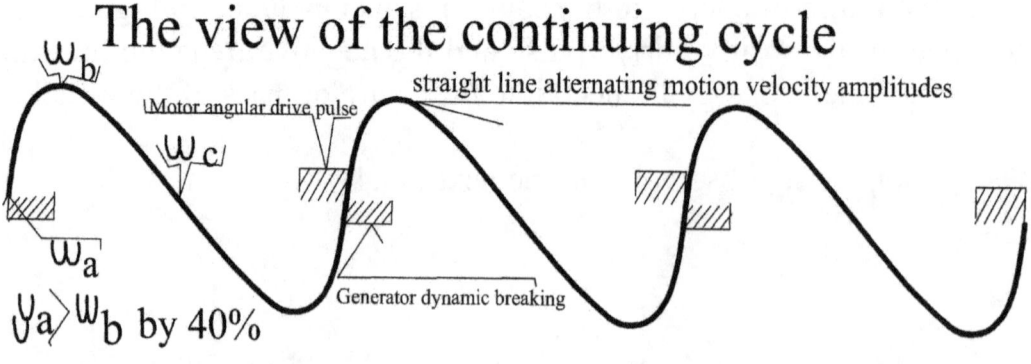

ω_b

Motor angular drive pulse

straight line alternating motion velocity amplitudes

ω_c

ω_a

Generator dynamic breaking

$\omega_a > \omega_b$ by 40%

Motor angular drive pulse magnitude minusGenerator dynamic breaking= Device Ke gain

The -..-..-.. lines are the angular velocities at each time position, black _____ lines are the reflecting (Newton's third law exception) straight line velocities at each position, the .... line are the instant force magnitude for each position and the hatched /////\\\\\ area is the impulse magnitude, the product of force and time. The angular velocity progressions ωa, ω_b and ω_c are indicated for a perfectly tuned resonant oscillator condition and is a perfect progression without any mechanical and magnetic friction. The rise of angular velocity from ω_b level to ω_c is normally present if the oscillator is not tuned to resonant condition. This is mathematically and graphically presented later.

The instant force magnitude for each time position is the slope of the change in velocity in respect to time duration, the slope of the solid _____ line in respect to time, this is the dv/dt acceleration gradient; when we multiply the dV/dt gradient with the flywheel1, 2 and motor generator mass we get the instant force magnitude at this time position.

In Fig.4 we are analysing the progression of the angular speed in respect to 1/5th time increments of the total cycle time. In view of the time domain contour Graph and the Fig.2 displacement domain graph, it can be postulated with certainty that the final resultant average net inertial propulsion force escalates proportionally with the cyclic input energy gradient and the instant force escalates in a quadratic progression with the magnitude of the oscillation frequency gradient. The thrust/impulse escalates proportionally with the angular speed gradient. This is easily observable because the area of average force times the time durations (the ///\\\ hatch areas), which is impulse, is larger in the forward direction than in the backward straight line motion direction and therefore proves the existence of a self-contained impulse.

The straight line displacement per each 1/4 crank turn is the area under the velocity curve in respect to time. The geometric dimension of the plot **proves again the validity of formula #13; that the mean value of the reciprocating impulses is the average of the angular speed per each quarter turn, P=m*r*1/2(ω_a+ω_b) multiplied by mass and by radius.** This additional proof is the proof by exhaustion. For reality-check of the presented impulse relationship view the pendulum test: www.mindbits.com/series/1278 section lesson 8.

The Next Fig.1b is a supplement to Fig.1 to identify the nature of the opposing impulses for each straight line starting and stopping motion.

131

Fig.1b is supplementing Fig.1 to present the impulses for every 1/4 turn of the cycle.

Fig.1b
Modified Scotch Yoke FM modulation

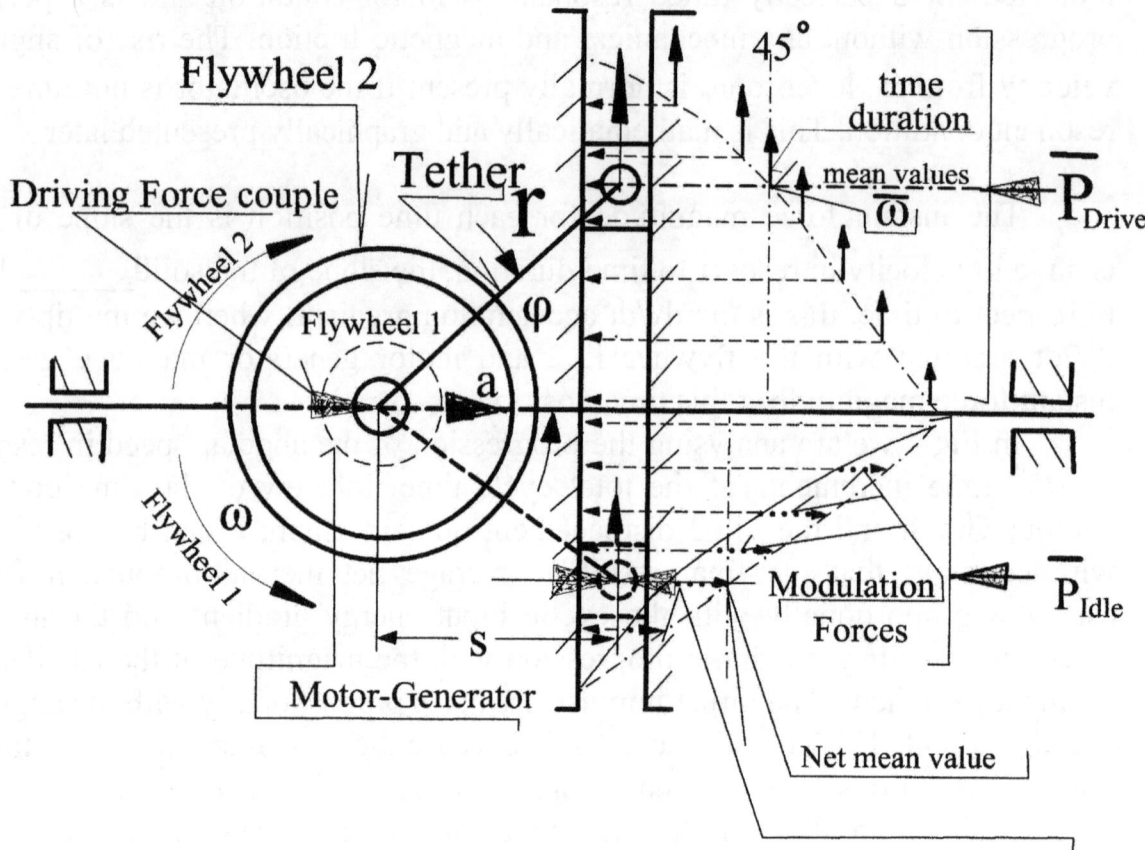

FM modulation opposing Force Couple

Newton's reaction to an action applies for the FM Modulation

New angular speed is introduced into Flywheel 1 without additional impulse

$$\overline{P}_{Drive} > \overline{P}_{Idle}$$

$\overline{P}_{Idle}$ = the impulses within the opposing starting and stopping motions

$$\overline{P}_{Drive} - \overline{P}_{Idle} \;\; is \;\; self\text{-}contained$$

Fig. 1c is helping to visualize to forces against each longitudinal section of the scotch yoke frame exerted by the flywheel tether. The horizontal dashed arrow length is representing the force magnitudes for each 18° flywheel 1 arc motion and the vertical arrows are the related time duration magnitudes, together they are representing the impulses. The vertical time arrows are in direction of the flywheel tether motion. The impulse drive and the opposing impulse is the net impulse, therefore this is **proving the existence of a net impulse drive**. Fig.1c depicts a method congruent with Newton's own pendulum derivation; the impulse sum from each section of a pendulum arc motion is the net impulse.

Fig.1c
Modified Scotch Yoke reciprocal Impulse

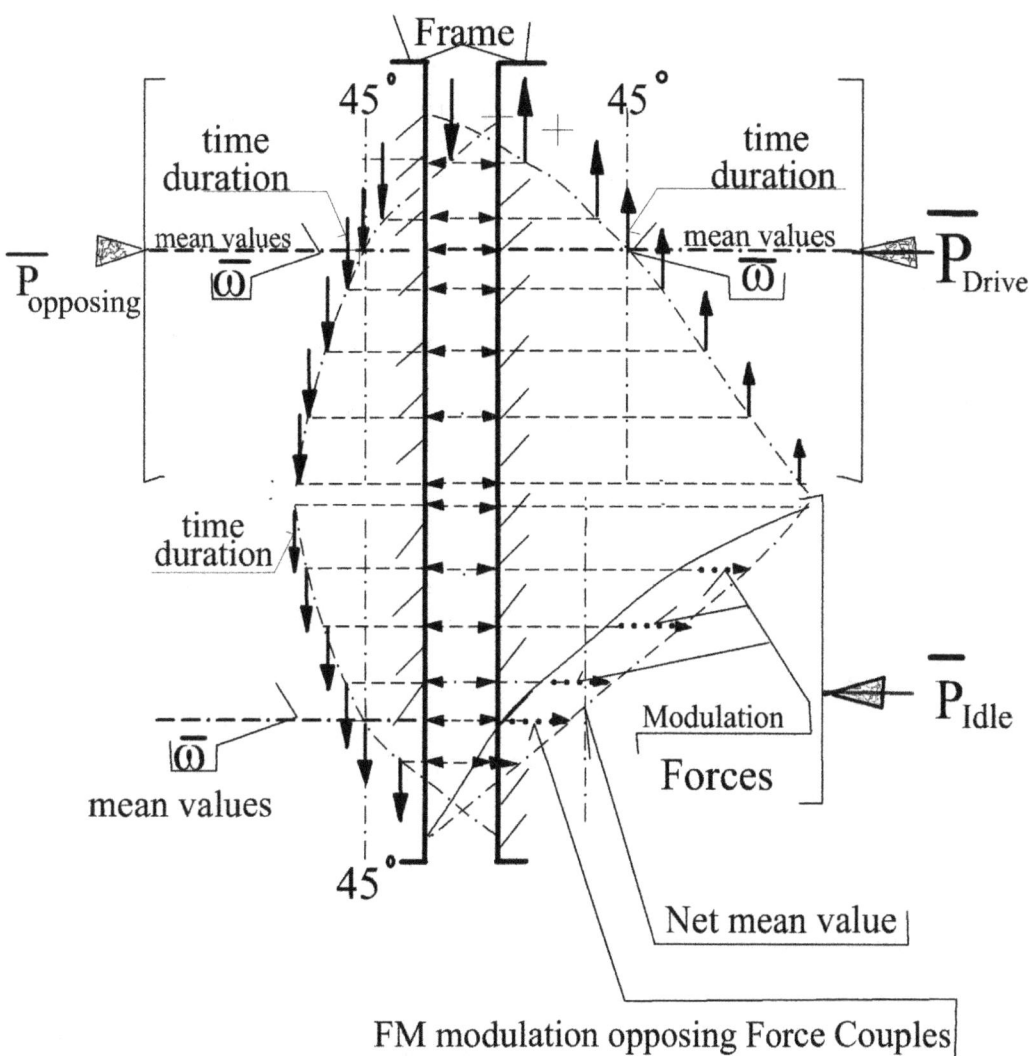

Fig.2b is supplementing of Fig.2 and proves the effect of velocity phase shift to real force. Here we observe that the Force is in the real domain and the velocity is in the imaginary domain, this proves that they are 90° phase shifted.

Next is the **Fig.1d** plot of the straight line motion energy flow in relation to the angular motion positions. Here we multiply the straight line motion incremental distance of the scotch yoke motion body in relation to the incremental angular position in 15° increments and multiply with the acceleration force applied for ωa= 100, 140, 160 and 180 rad/second; this is the energy flow per this 15° section. The energy flow progression indicates an avalache flow centerd at approcimately 40° agular position!

The Fig.1d graph indicates / proofs that there is no energy restriction for the maximum energy flow magnitude, only the mechanical construction material stencil strenght consideration is the underlying limitation!

Fig.1d
Apparent straight line motion energy flow in relation to angular rotation

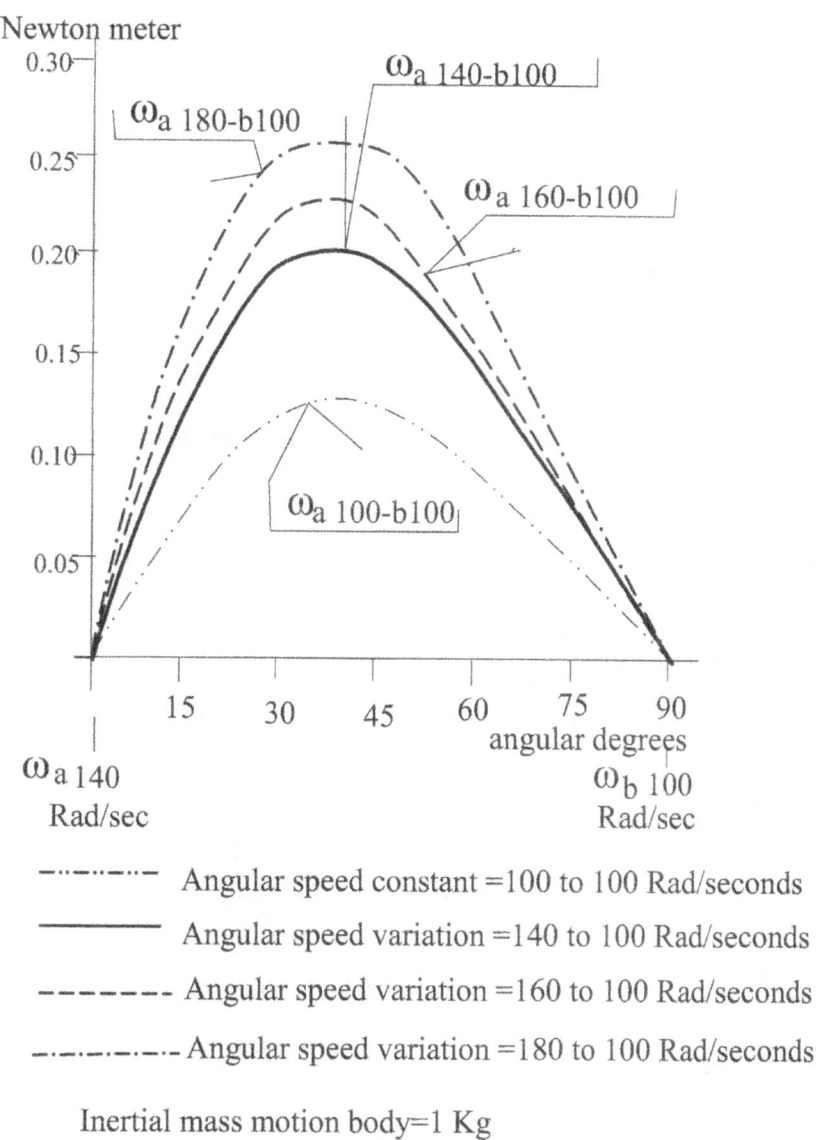

Newton meter

ω_a 180-b100

ω_a 140-b100

ω_a 160-b100

ω_a 100-b100

angular degrees

ω_a 140
Rad/sec

ω_b 100
Rad/sec

—··—··— Angular speed constant =100 to 100 Rad/seconds

———— Angular speed variation =140 to 100 Rad/seconds

—————— Angular speed variation =160 to 100 Rad/seconds

—·—·—·— Angular speed variation =180 to 100 Rad/seconds

Inertial mass motion body=1 Kg

Fig.2b
1/4 turn mean value Velocity plot

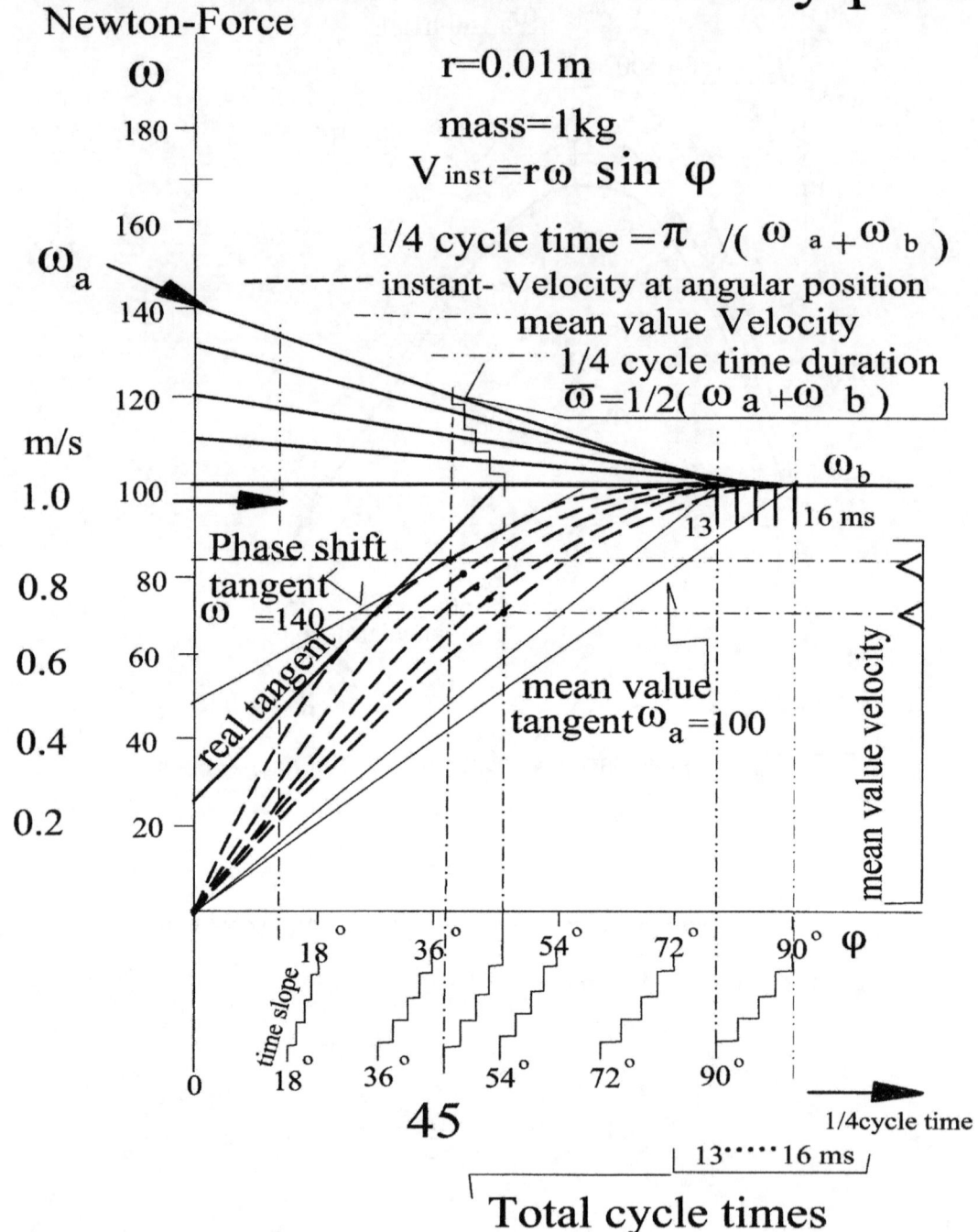

The Proof of the reactionless FM Modulation Ke Accumulation Phase

During the accumulation phase, rotational kinetic energy is accumulated into the impact rotor (Flywheel 1), by mutual reciprocal and singular exertion against the freewheeling rotational reluctance of the flywheel (Flywheel 2, the dynamic backrest). The action of the accumulation of the rotational kinetic energy is reaction-less in relation to the device mass due to the reciprocal action presented in formula#8. The reciprocal action is performed, because the new induced rotational kinetic energy was developed mutually reciprocally and singular against the freewheeling inertial mass reluctance of the backrest flywheel (Flywheel 2). The declining slope of the straight line velocity of the flywheel assembly motion body, in respect to the distance moved by the motion body is causing the conservation of straight line motion kinetic energy into rotational kinetic energy of the impact rotor (Flywheel 1). The declining slope is illustrated by the Cartesian coordinate in the drawing Fig.2c. It has the effect of removing the straight line kinetic energy from the motion body mass, even when new straight line kinetic energy is introduced and it is feeding the kinetic energy into the impact rotor (Flywheel 1), which reduces the straight line kinetic flywheel energy to zero at the TDC position. Such a negative slope of the velocity is the characteristic of this combined motion of rotation and straight line motion. It can be further concluded that the action of the accumulation phase has no net kinetic effect on the device mass, due to the equal action and reaction of all the straight line forces at play and the fact that all straight line kinetic energy has been fed into the impact rotor at the end of the accumulation phase; the accumulation phase is in fact congruent with the null effect of the throw of an inertial mass within a vehicle. In conclusion, the accumulation phase is complying with and is working with the principle of conservation of kinetic energy and the conservation of momentum. It is furthermore possible to place the accumulation Phase in congruence with dynamic breaking where the accumulator is the impact rotor (flywheel 1) of the rotational to reciprocating transmission.

Ref. Kurt Gieck Engineering Formulas P.10; and epi-eng.com.

The next diagram, a Vector plot, is presented to further explain the Accumulation Phase Force distribution with Vectors; here we proof again that the "Accumulation phase" is accomplishing the frequency modulation of the modified scotch yoke mechanical oscillator without impinging new impulses onto the isolated system of the device. Frequency modulation means that the transitions from positive to negative energy pulses occur at variable time durations.

Fig.2c

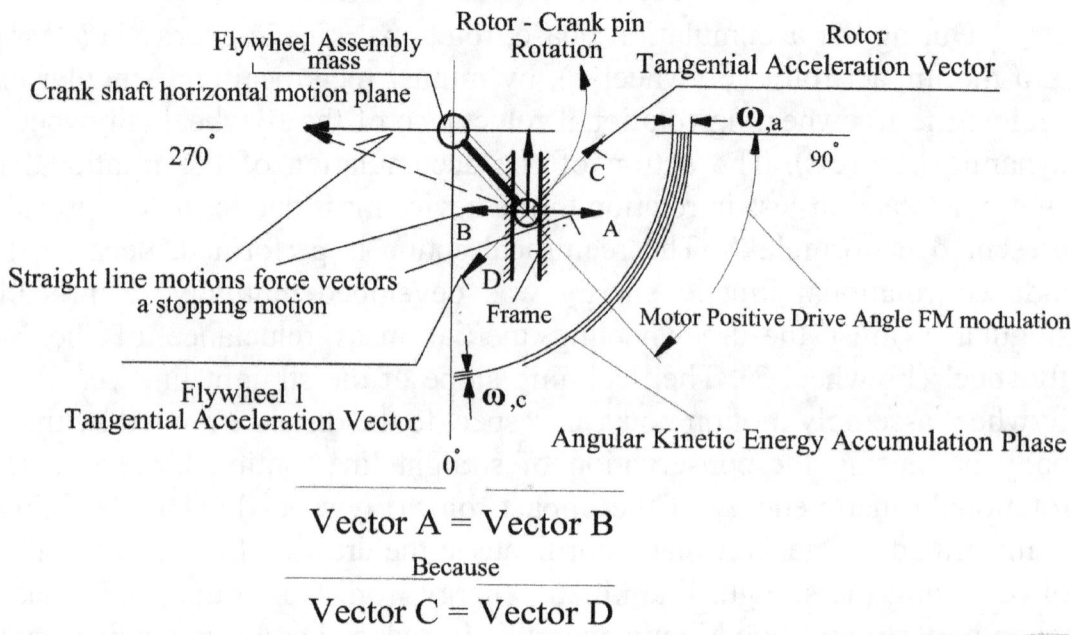

Vector A = Vector B

Because

Vector C = Vector D

Proof: **Vector A is equal to Vector B because the tangential acceleration Vector C is equal to the Flywheel acceleration Vector D. Therefore, the modulation is a net Zero Force and the device is immune to the FM modulation. The FM modulation is in fact, congruent with an inertial mass throw within the confines of vehicle!** Important: The ¼ turn energy accumulation phase time duration is in congruence with straight line motion time formula #1.7; however, here it is in angular form: the higher the moment of inertia the higher the time duration, or the higher the torque the shorter is the time duration in a diminishing returns progression. The torque is the drive energy divided by the angular displacement and is the quantity we are able to control #1.7: $t_{time} = (2S_{distance} m / F_{force})^{1/2}$;

$$t_{\omega c \omega a} = \pi_{angular, ¼distance} \tfrac{1}{2} / \omega_c + (\pi_{angular, ¼distance} I / \tau_{torque})^{1/2}$$

Ref. Kurt Gieck Engineering Formulas P.10; and epi-eng.com.

Here we arrive at the completion of the formal proof of Inertial Propulsion. We presented four mechanical constructs, the Ice Skaters, the modified Trebuchet, the oblique pendulums and the modified forced oscillation Scots Yoke; we presented a graphical vector-kinematic plan view proof and a numerical algorithm proof positive, all capable of proof positive Inertial Propulsion and three proofs positive pendulum test of these devices; this is a proof by total exhaustion.

Here is the author's **I**nertial **P**ropulsion design software program listing available with the purchase of this book for maximizing the inertial mass configuration and delivering a better understanding of the internal processes. The Programming language is basically Fortran.

```
[top]
rem 5/12/2015
cls
print "|-------------------------------------------------------------------
-------------------------------------------------------------------|"
print "|    A Computer Program for calculating the Fig.1, Fig.2 and
Fig.9 Scotch Yoke mechanism's inertial Forces, Impulses, cycle time,
and Velocities        |"
print "|    The Scotch Yoke mechanism is delivering a true sinusoidal
oscillating alternating straight line motion reflected from a crank/cam
orbital motion        |"
print "|The mechanical principle of the scotch yoke relates to the
centripetal force applying to the orbital motion of an inertial mass
F=mV^2/R=mW^2*amplitude   |"
print "|The program accumulates the net angular inertial mass motion
impulse and the net Velocity response of the total scotch yoke inertial
mass system per cycle|"
print "|                              This Program was authored by
Gottfried J. Gutsche                              |"
print "|-------------------------------------------------------------------
-------------------------------------------------------------------|"
print " "
[A] input " Specify Scotch Yoke Crank Pin orbital angular speed Wa
at Top Dead Center (TDC, the motionless point) in 2pi/time,
RAD/second="; WTDC
if WTDC=0 then [e] else [B]
[e] print "| input "; WTDC; " not valid, enter a valid W number "
goto [A]
[B] input " Specify Scotch Yoke Crank Pin orbital angular speed Wb
```

at Top Dead Center + 1/4 additional turn, wherein Wb<Wa, in 2pi/time, RAD/second="; WTDC90

if WTDC90=0 then [e] else [R]

[e1] print "| input "; WTDC90; " not valid, enter a valid W number "

goto [B]

[R] input " Specify crank pin radius distance in meter ="; R

if R=0 then [e2] else [dS]

[e2] print " input "; R; " not a valid R length "

goto [R]

[dS] input " Specify derivative angular crank pin calculation step increments dS in angular degrees ="; dS

if dS=0 then [e3] else [Kg]

[e3] print " input "; dS; " not a valid dS angle step"

goto [dS]

[Kg] input " Specify Scotch Yoke motion body inertial mass in Kg="; mass

if mass =0 then goto [e4] else goto [flywheel1]

[e4] print " input "; Kg; " not a valid response ";

goto [Kg]

[flywheel1] input " Specify Flywheel #1 mass in Kg="; flymass

if flymass =0 then goto [e5] else goto [flywheelrad]

[e5] print " input "; flymass; " not a valid response ";

goto [flywheel1]

[flywheelrad] input " Specify Flywheel 1 radius in meters="; flyrad

if flyrad =0 then goto [e6] else goto [totalmass]

[e6] print " input "; flyrad; " not a valid response ";

goto [flywheelrad]

[totalmass] input " Specify total mass of the mechanisem in Kg="; totalmass

if totalmass =0 then goto [e7] else goto [reset]

[e7] print " input "; totalmass; " not a valid response ";

goto [flywheel1]

[reset]

de=0.000

Waverage=0.000

```
Vnew=0.000
VA=0.000
SA=0.000
PA=0.000
dt=0.000
ST=0.000
dST=0.00
TA=0.000
EA=0.000
Faverage=0.000
x=0.0000
Kyoke=0.0000
fly1e=0.000
fly1e90=0.000
Fmax=0.000
stepup=1
[again]
Print " "
print " Calculating Run# is now ="; stepup; "th run"
string$= " Ke=EA; accumulate energy steps de into accumulator EA;
EA=de+EA ; Run#= "; stepup
if stepup < 2 then print string$
EA=de+EA
string$= " average linear velocity="; Vaverage; " meter/second"
if stepup < 2 then print string$
string$=" accumulating TA; TA=TA+dt; TA= "; TA; " dt="; dt; "
second"
if stepup < 2 then print string$
TA=TA+dt
string$= " TA="; TA; " dt="; dt; " second"
if stepup < 2 then print string$
string$=" SA="; SA ;" SA+dS/2="; (SA+dS/2); " degree at current
angular position"
if stepup < 2 then print string$
string$=" WTDC="; WTDC; " WTDC90="; WTDC90
```

```
if stepup < 2 then print string$
x=(SA+dS/2)
y=x*3.141/180
string$= " COS((SA+dS/2*)3.14159/180)="; COS(y)
if stepup < 2 then print string$
string$= "calculate change of angular speed Kyoke per delta angular
displacement dS; formula= Kyoke=WTDC90-WTDC/90"
if stepup < 2 then print string$
Kyoke=(WTDC-WTDC90)/90
string$ = " angular change slope Kyoke in dW/degree="; Kyoke; "
dW/degrees"
if stepup < 2 then print string$
string$=" calculate the straight line distance from formula:
ST=R*COS(SA+dS/2)"
if stepup < 2 then print string$
string$= " cos of angular degree at SA position= COS"; SA+dS/2
if stepup < 2 then print string$
STo=ST
rem print " STo= "; STo; " meter";
x=(SA+dS/2)
rem print " x="; x; " y="; y
y=(x*3.1415/180)
print " calculating the straight line motion distance with:
COS((SA+dS/2*)3.14159/180)*R "
ST=COS(y)*R
dST=ST-STo
print " at straight line distance="; ST; " meter"; " change of straight
line motion distance, dST= "; dST; " meter"
string$=" calculate average Force per distance accumulation SA
degree using Kyoke="; Kyoke
if stepup < 2 then print string$
Waverage=(Kyoke*(90-SA-dS/2))+WTDC90
print " at SA degree position "; SA+dS/2; " degrees; angular speed is
at ="; Waverage; " RAD/second"
print " calculate now average straight line Force Faverage at angular
```

position SA+dS/2, formula:
Faverage=mass*Waverage*Waverage*ST"
print " calculationg now average force Faverage at angular
position="; SA+dS/2; " degrees"
Faverage=mass*Waverage*Waverage*ST
if stepup < 2 then Fmax=Faverage
de=dST*Faverage
print " straight line force="; Faverage; " Newton"; " angular turn
distanse SA+dS/2= "; SA+dS/2; " degrees"; " linear distance"; ST; "
meters, dEnergy de="; de; "Newton meter"
rem " calculating de"
REM print " inear Force average at SA+dS/2 ="; (SA+dS/2); " force
average="; Faverage; "Newton"
rem " calculate Vaverage= Waverage*R*SIN(y)"
rem print " SIN(y=)"; SIN(y); " Waverage="; Waverage; " R="; R
x= SIN(y)
rem print " SIN(y=)"; x
Vaverage=x*Waverage*R
Print " at "; SA+dS/2; " degrees angular displacement the straight
line motion velocity is; Vaverage= "; Vaverage; " meter / second"
print " dST= "; dST
dt=dST/Vaverage
print "angular step dS= "; dS; " degree; Waverage= "; Waverage; "
RAD/second"
dt=0.01745*dS/Waverage
print "delta time for the dS displacement, dt= "; dt; " second"
SAo=SA
print " now accumulating the angular position SA in degrees"
SA=SA+dS
PA=PA+(dt*Faverage)
stepup=stepup+1
print " Ke="; EA; " degrees ";
print "angular distance accumulated="; SA; " degree"; " accumulated
linear distance="; ST; " meter "; ", now going to calculating next
Run# "; stepup; " run"

if SA > 91 then [display]
if SA = 90 then [display]
goto [again]
[display]
print " "
print " FINAL RESULT: Apparent energy per TDC to TDC90 turn cycle=";EA;" Newton meter"
print " FINAL RESULT: Real straight line impulse sum per TDC to TDC+90 turn cycle P="; PA; " Newton seconds"
print " FINAL RESULT: TDC to TDC90 (90 degree) rotation cycle time duration="; TA; " seconds"
print " FINAL RESULT: Final Scotch Yoke internal motion mass ="; mass; " Kg straight line speed amplitude="; Vaverage; " meter/second"
fly1e=WTDC*WTDC*flymass*flyrad*flyrad/4
print " Flywheel 1 initial angular potential energy at TDC= ";fly1e; " Newton meter"
fly1e90=WTDC90*WTDC90*flymass*flyrad*flyrad/4
print " Flywheel 1 angular kinetic energy at TDC+90 degrees= "; fly1e90; " Newton Meter"
dMKe=(PA*PA-(WTDC90*R*mass))/(2*totalmass)
if (fly1e-fly1e90) < dMKe then [e8] else [e9]
[e8] print " not sufficient flywheel 1 inertia!"
input " type letter a to run program again ="; x$
if x$ ="a" then goto [top]
if x$ ="A" then goto [top]
end
[e9] dV=(PA-(Vaverage*mass))/totalmass
Print " Scotch Yoke aggregate mass = "; totalmass; " Kg straight line velocity gain per cycle, dV= "; dV; " meter / second"
x=((2*TA)+(4*3.1415/WTDC90))
print " Total 360 degree turn cycle time duration ="; x; " seconds"
print " Continuous Thrust of the device="; (PA-(mass*Vaverage))/(x*0.102); " Gramm "
print " Scotch Yoke aggregate mass ="; totalmass; " Kg straight line

velocity gain per hour ="; ((3600*dV)/x); " meter / second"
print " Mechanism internal Peak Force ="; Fmax; " Newton"
print " Total Net straight line impulse= "; (PA-(mass*Vaverage)); " Newton second"
print " flywheel 1 angular kinetic energy safety margin (the -energy returned to the source) at TDC ="; (fly1e-fly1e90); " Newton meter"
x= 1.2*(fly1e+fly1e90)*(WTDC+WTDC90)/2
print " Drive Motor-Generator shaft minimum output power capacity ="; x; " Newton meter/ second (watt)"
input " type letter a to run program again ="; x$
if x$ ="a" then goto [top]
if x$ ="A" then goto [top]
end

This is the final analysis report print out at the end of the program run for a very large IP acceleration exceeding the pull of the Earth: **ω,a TDC=28501 1/sec ; ω,b TDC+90=1500 1/sec.; crank radious=0.021meter; Flywheel 1 ma**ss=:2.3 Kg; Flywheel1 radius=0.06 meter; total system mass**=5.2 Kg**

FINAL RESULT: Apparent energy per TDC to TDC90 turn cycle=6468.03302 Newton meter

 FINAL RESULT: Real straight line impulse sum per TDC to TDC+90 turn cycle P=92.1142695 Newton seconds

 FINAL RESULT: TDC to TDC90 (90 degree) rotation cycle time duration=0.10638499e-2 seconds

 FINAL RESULT: Final Scotch Yoke internal motion mass =2.35 Kg straight line speed amplitude=6.27600257 meter/second

 Flywheel 1 initial angular potential energy at TDC= 4386.15 Newton meter

 Flywheel 1 angular kinetic energy at TDC+90 degrees= 12.15 Newton Meter

 Scotch Yoke aggregate mass = 5.2 Kg straight line velocity gain per cycle, dV= **14.8780122** meter / second^2;

Total 360 degree turn cycle time duration =0.85901033e-1 seconds

Continuous Thrust Force pulses of the device=8829.77619 Gramm

Scotch Yoke aggregate mass =5.2 Kg straight line velocity gain per hour =623518.042 meter / second

Mechanism internal Peak Force =358392.58 Newton

Total Net straight line impulse= 77.3656635 Newton second

flywheel 1 angular kinetic energy safety margin (the -energy returned to the source) at TDC =4374.0 Newton meter

Drive Motor-Generator shaft minimum output power capacity requirement =6597450.0 Newton meter/ second (watt)

type letter a to run program again =

Accordingly a=14.8m/sec^2 is sufficient to escape the gravitational pull G=9.8m/s$^{2.}$

Optimum angular Speed TDC to TDC+90 spread selection

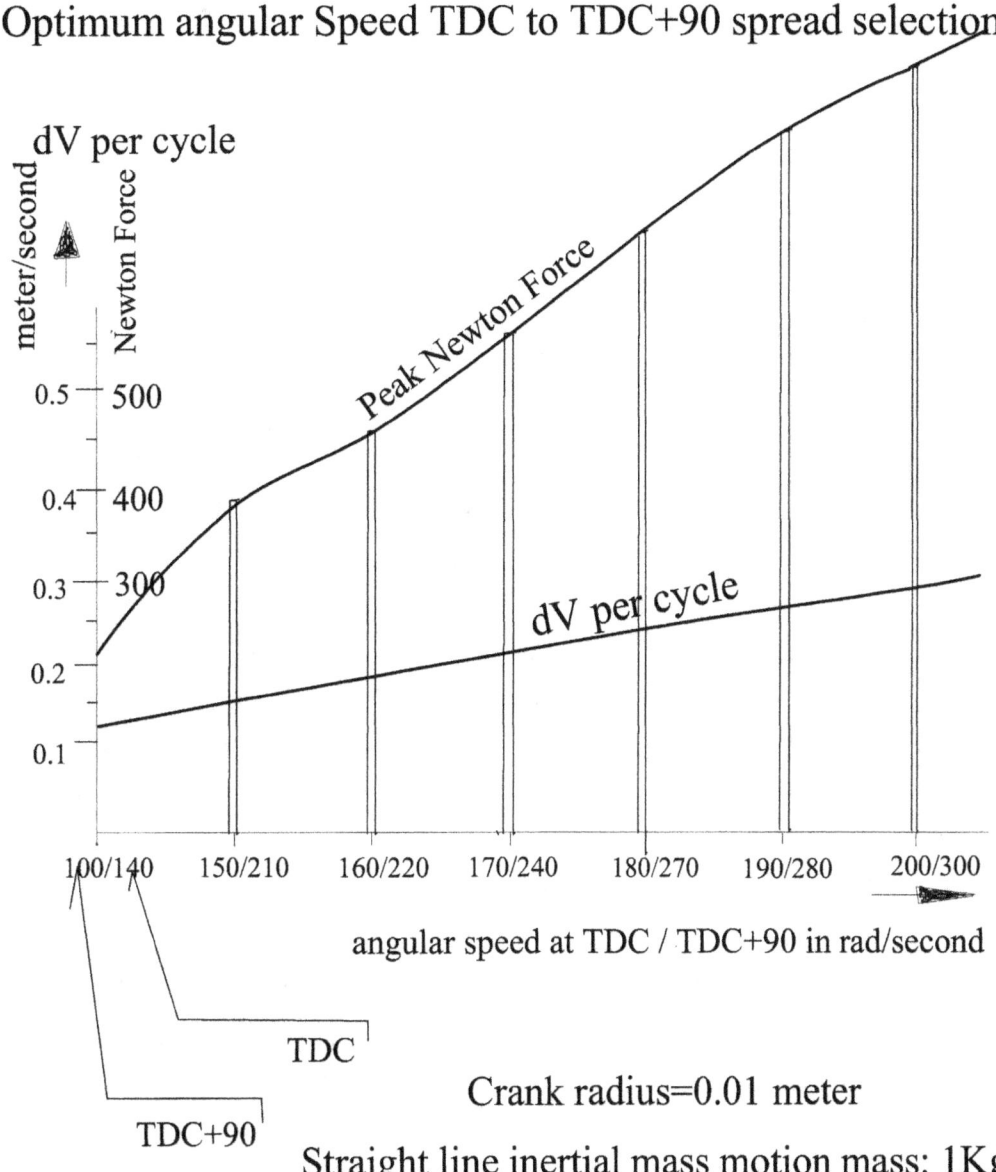

Crank radius=0.01 meter

Straight line inertial mass motion mass: 1Kg

Next is the plot of the Program output from ω_a=100 to ω_a=200 rad/second, while ω_b =100rad/second is staying steady:

Impulse and Velocity change in relation to Angular speed at TDC

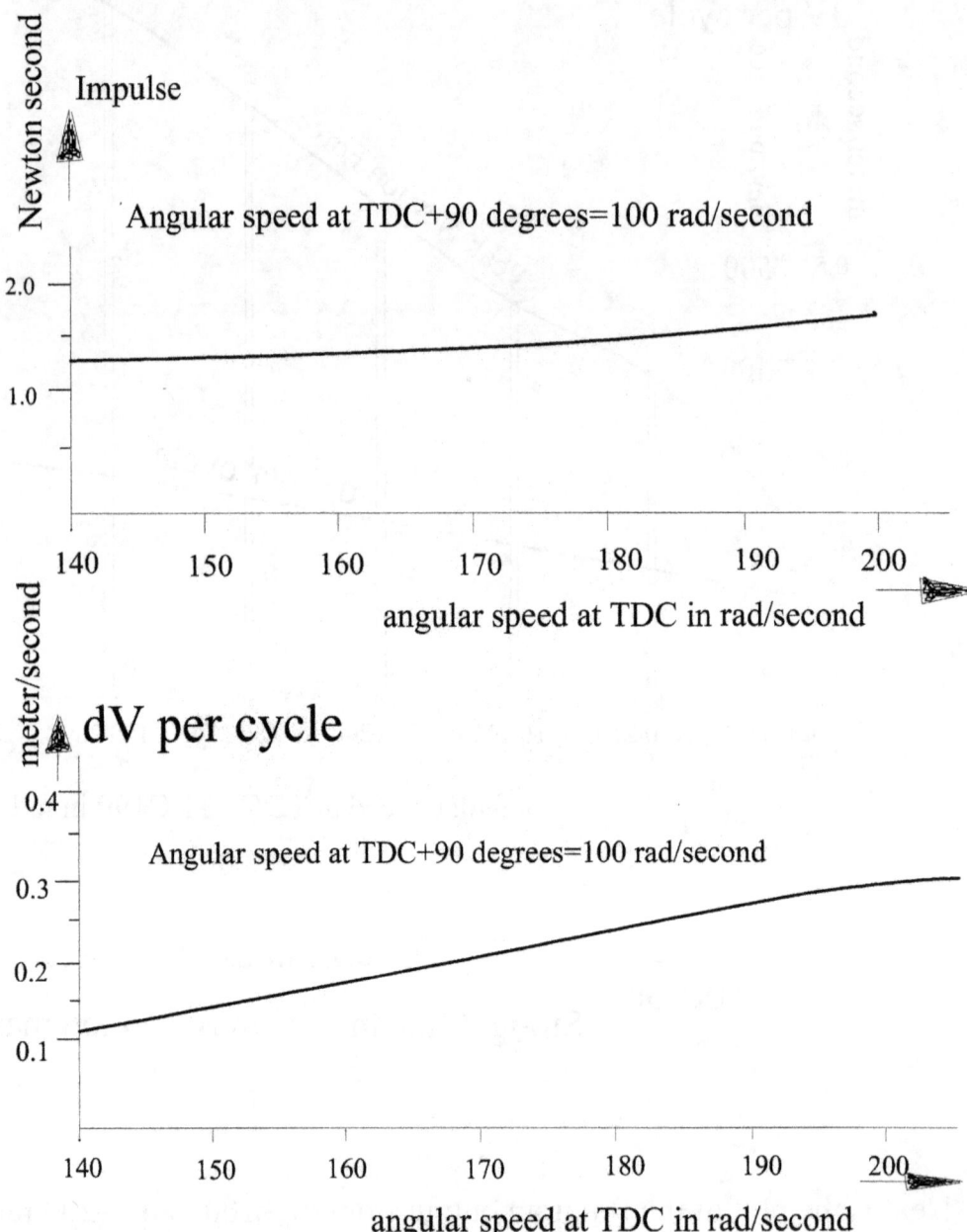

Newton second

Impulse

Angular speed at TDC+90 degrees=100 rad/second

2.0

1.0

140   150   160   170   180   190   200

angular speed at TDC in rad/second

meter/second

dV per cycle

0.4

Angular speed at TDC+90 degrees=100 rad/second

0.3

0.2

0.1

140   150   160   170   180   190   200

angular speed at TDC in rad/second

Math footprint for calculating the non-uniform rotational to straight line coupled motion

Referencing Fig.2, Fig.2b, Fig.2.c and Fig.4. The next Fig.5 is the analysis of the angular velocity progression in relation to the impulse and kinetic energy values. Fig.4 is the complete motion progression graph of the presented continuous cycling device.

Rotor Angular Velocity at 0°, the start of straight line motion = ω,a

Rotor Angular Velocity at 90°, the amplitude of straight line motion = ω,b

$$\omega,a > \omega,b$$

Average (Mean Value) Angular velocity for the ¼ Rotor turn:

mean, value $= (\frac{1}{2}(\omega,a + \omega,b))$

Squaring the mean angular velocity we get $(\frac{1}{2}(\omega,a + \omega,b))^2$

The averaging factor for ¼ turn is: $2/\pi$

Therefore, the Average force for non-uniform Rotational to Straight line Coupled motion for a drive phase is:

Force, average,¼ turn, non uniform =
mass, straight line, displacement*Radius*$(\frac{1}{2}(\omega,_a+\omega,_b))^2\ 2/\pi$

Therefore by multiplying Force, average, ¼turn, non uniform * Time, duration, $2/\pi(\frac{1}{2}(\omega,a +\omega,b))$ turn; we get Impulse:

Time, duration,¼, turn $=\pi/(2(\frac{1}{2}(\omega,a +\omega,b)))$

Impulse, average,¼ turn $=\text{mass}_{\text{straight line, displacement}} * \text{Radius}_{\text{crank}}$ $*(\frac{1}{2}(\omega,a+\omega,b))$

Average angular velocity for rest of the ¾ rotor turn is a $\omega,_b$ progression rising up by the multiplying $\omega,_b$ with constant $C2$. Then slowing back down by multiplying $\omega,bC2$ with constant $C1$ because of the kinetic energy flow into and out of the straight line mass motion and friction losses. This principle is presented in the next Fig.5:

Fig.5

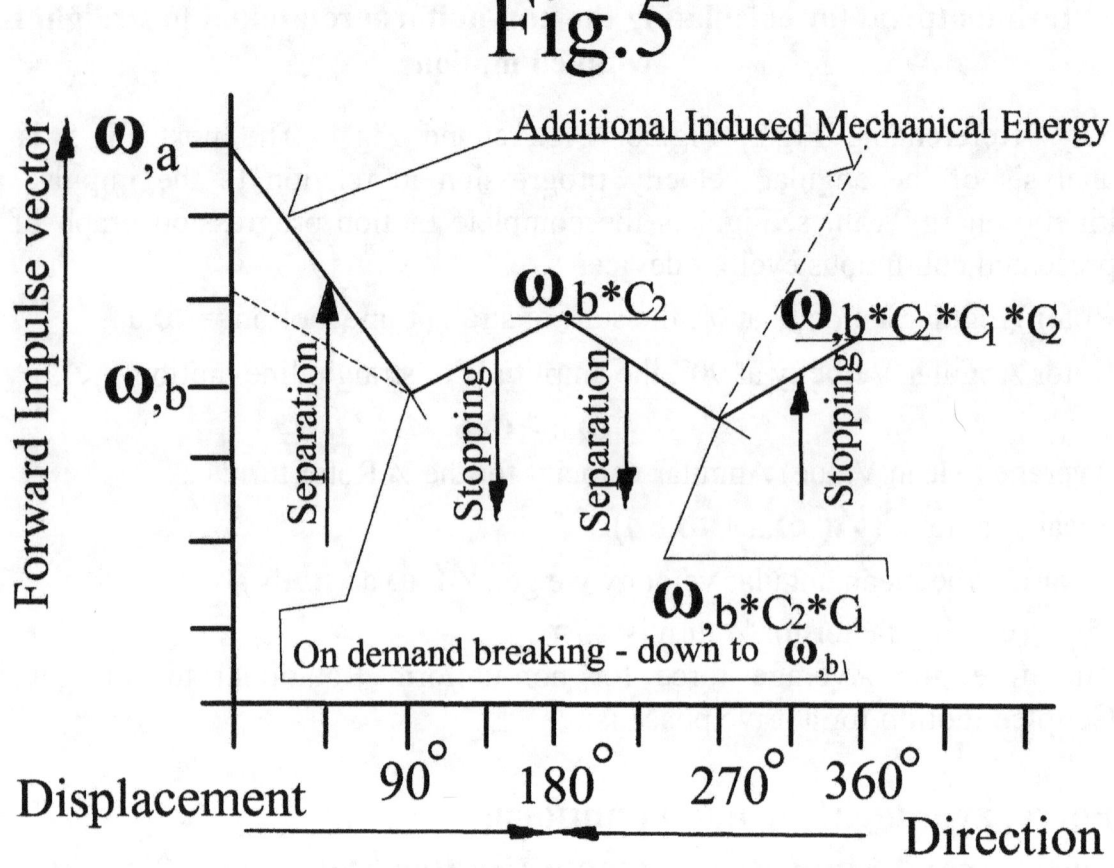

The Fig.5 reveals that the drive phase impulse having a larger mechanical energy potential is $(\frac{1}{2}(\omega,a +\omega,b))$average progression and is opposed by one Idle phase impulse having a $(\frac{1}{2}(\omega,b +\omega,bC_2))$ average progression. The ω,bC_2 accounts for the kinetic energy flow from the stopping of the straight line displacing mass during the Idle Phase presented in Fig.5. Thereby, the three impulses progressing for each 1/4 turn after the drive phase must be alternately subtracted and added to arrive at the exact resultant internal self-contained impulse. The sum of impulses during the idle phase:

impulse, idle=mass *radius*$(-(\frac{1}{2}(\omega,b+\omega,b\ C_2))$-$(\frac{1}{2}(\omega,b+\omega,b*C_2*C_1)+(\frac{1}{2}(\omega,b+\omega,b\ C_2^{2}*C_1))$

Refined to: Impulse, idle=mass *radius*$(\frac{1}{2}\omega,b(-1-C_2-1-C_2C_1+1+C_2^{2}C_1))$
Drive phase and Idle Phase will be defined later. The idle phase impulses from 90° to 360° can be algebraically solved to the mathematically exact:

Impulse,idle=mass*radius*1/2ω,b(-1-C2-C2C1+C2²C1)

C3=(-1-C2-C2C1+C2²C1)

The Net Impulse is: **Impulse, net= mass, straight line, flywheel, displace * Radius,crank *½(ω,a+ω,b(1+C3))**

The Self-contained internal impulse is divided by the total cycle time duration to arrive at the **NET** Internal self-contained motivating force, because the total cycletime is diluting the generated impulse. Accordingly, the total Cycle Time is:

¼cycle time **t1=π/(ω,a +ω,b))**, plus one ¼cycle time **t2=π/(ω,b +ω,b * C2))** plus **t3=π/(ω,b * C2+ω,b * C2*C1))**.

If no new mechanical energy is induced during **t4** then we add:

time **t4=π/(ω,b*C2*C1+ω,b* C2*C1*C2)**.

If **new** mechanical energy is induced then ω,b * C2*C1 is boosted up to ω,a then we add: **t4=π /(ω,b*C2*C1+ ω,a)**

Thereby: The effective internal motivating force, the Internal

Propulsion Formula #14is:**Force, internal, self-contained=mass, flywheel*Radius, orbit*½(ω,a+ω,b(1+C3))/(t1+t2+t3+t4)**

Thereby, the self-contained Inertial Propulsion motivating energy is:

Ek=mass,flywheel*Radius²,orbit*½((ω_a+ω,b)(1+C3))/(t1+t2+t3+t4)

phase shift triangle graph used in electrodynamics phase shift Physics. It further proves the congruence of inertial mass motion with electrodynamics:

Functional Elements of the Inertial Propulsion Drive

The described combined rotational and straight line motion effort inertial drive has seven main functional elements:

1. A pair of flywheels (flywheel #1,#2) for providing a rotational inertial reluctance backrest to produce a reaction-less rotational force impulse. The pairs of flywheels move in alternating straight line reciprocal motion in direction of vehicular travel, and have parallel axial orientation, opposing rotations, equal peak straight line motion velocity amplitudes, equal straight line stroke length and differential straight line reciprocal motion cycle times. Each flywheel has complete freewheeling freedom of rotation in relation to the propulsion device.

2. An impact rotor (flywheel #1) in axial alignment with each flywheel for the purpose of temporarily accumulation and storing rotational kinetic energy with angular exertions against the backrest flywheel #2 to be used for the propulsion of the device; this flywheel to rotor dynamic angular exertion is independent of exertions against the vehicle and is **the ESSENCE** of the present invention.

3. A mechanical energy generator in the form of a motor-generator engaging with each flywheel and is mutual and reciprocally engaging with each impact rotor (#1). The motor-generator has the purpose of energizing the impact rotor with rotational kinetic energy, while using the flywheel #2) as the inertial backrest. The flywheel, the impact rotor and the motor-generator are assembled into an integral assembly. Without the flywheel #2 there is no self-contained inertial propulsion impulse possible.

4. A transmission for converting the rotation of the impact rotor into reciprocating straight line motion of the flywheel assemblies. The transmission, therefore, can be called a rotational-to-reciprocation transmission. For the purpose of mathematical simplicity, a complimentary cam and cam followers are used as a rotational-to-reciprocation transmission for all the following propulsion discussions, because of the simplicity of the straight line rise and fall of the straight line velocity in relation with the rotation of the impact rotor. This type of motion is also referred to as a saw-tooth motion.

5. A pair of straight line guides for guiding the flywheel assemblies in a straight line reciprocal motion.

6. A reciprocal touch friction break shoe or an electro-mechanical dynamic break for removing excess from flywheel #2 angular velocity mutually and reciprocally between flywheel #2 and the frame. The touch friction break does not interfere with the propulsion of the device because of its mutual and reciprocal operation; it also represents the simplest form of the device.

7. A supporting frame, for the purpose of supporting items 1 to 6.

Description of the combined motions Propulsion Cycle

The combined rotational and straight line motion, inertial propulsion, is accomplished with a four phase process. Each phase is a quarter turn of the impact rotor. The impact rotor rotation is used as a measure of reference, because of the workable characteristic of the angular motion position as a reference and because of the variable character of the cycle time duration. The impact rotor direction of rotation is counter-clockwise; the rotationof the flywheel is clockwise.

1. Accumulation Phase

The Accumulation Phase accomplishes the accumulation of rotational kinetic energy into the impact rotor (flywheel #1) by mutual rotational reciprocal force exertion against the reluctance of a flywheel #2, by the motive force of the motor-generator. The accumulation phase increases the angular velocity of the impact rotor. The utilization of the motor-generator is 1/8 of the total capacity of the motor, because of the ¼ turn of the drive phase and the reciprocal exertion between the flywheel and the impact rotor, which distributes kinetic energy into both the impact rotor (#1) and the flywheel #2). Real power is 1/8 of nominal rated power.

2. Drive Phase

The Drive Phase release of the rotational kinetic energy is accumulated in the impact rotor and into the straight line inertial kinetic energy of the flywheel and into the straight line inertial kinetic energy of the propulsion device by mutual and reciprocal inertial mass separation based on formula #8 and example #5.

3. Rotor Break Phase

Is causing the removal of excess (unused) rotational kinetic energy from the impact rotor (#1) to accomplish the angular velocity $\omega_{,b}$. The impact rotor break phase is an on-demand function, which depends on the relative resistance of the device to motion. The impact rotor break phase occurs during the end of the drive phase, **before the straight line velocity amplitude is reached and is therefore partially a simulated expulsion of mass**. The break phase is a complex vector force causing an impact rotor de-acceleration which is exerted mutually reciprocally between the impact rotor and the flywheel 2. The intensity of the break phase, in combination with the drive phase, also determines the overall gain/variance of the angular velocity of the impact rotor.

4. Idle phase

When no new energy is induced into the impact rotor, the stored rotational kinetic energy of the impact rotor will flow alternately into the straight line kinetic energy of the flywheel assembly and back into the impact rotor through the motion of the rotational-to-reciprocating transmission.

The straight line reciprocating motion of the flywheel assemblies and the alternating acceleration and de-acceleration of the impact rotor is an alternating feedback loop. The straight line reciprocating flywheel motion has an equal peak velocity and the straight line acceleration and de-acceleration forces of the two flywheel assemblies are in reciprocal equilibrium. Therefore, no motion or vibration forces act onto the device. The idle motion frequency is preferably the maximum allowable motion frequency of the employed mechanical design.

The accumulation phase and drive phase each represent one quarter turn of the impact rotor, for a total of one half cycle. The second half of the total propulsion cycle is composed of two consecutive idle phases one quarter turn each.

Important: The idle phase, when considering an ideal performance condition, must be viewed applying to Newton's Second Law: "Without the presents of any new forces no loss of momentum will occur during the Idle Phase". The ideal condition of no loss of momentum is a difficult thing to accomplish, because, the driving element is a motor generator having a magnetic drag. This drag is expressed as a Drag Constant C having a dimension of Newton seconds/meter wherein the drag force is:

$$F = cV \text{ in dimension of } N_{ewton}.$$

This means the higher the velocity the higher the **DRAG FORCE**, accordingly the impact rotor angular velocity ω is limited when using a Motor-Generator with a Magnetized rotor. For an ideal IP device an "Iron-Less Rotor technology" is required; current ironless rotor technology has a limitation in motor torque.

Arguments about the shifting Reference Frame

An argument contrary to IP is raised by some publications regarding the shifting-accelerating reference frame (the accelerated frame of reference). "How is an internal force able to perform this acceleration in view of the reference frame?"

When using the law of mean value, the angular velocity $\omega_{,b}$ multiplied by the crank radius is the straight line motion flywheel 1+2 assembly inertial mass velocity gain, then divided by the cycle time is the mean value straight line displacement force applied by the cyclic motion. The motivating force is then the same as the resultant force from the subsequent reactive impulse. "Is Newton's equal reaction to an action shutting IP down?". "NO, this argument is not shutting down the inertial propulsion show", because the presented inertial propulsion is using a non-harmonious cyclic mass motion. This motion is transmitted through a transmission ratio having a frequency modulated variable cycle time, non-uniform crank rotor circular motion and an average angular velocity which keeps the cyclic velocity gain of the sinusoidal straight line displacement magnitude IN-variable and the cycle time VARIABLE. Therefore, the propulsion force is a quadratic progressing variable parameter outperforming -outrunning the non-uniform changes in cycle time. The straight line displacement of the flywheel 1+2 assembly motion velocity gain is invariably repeating. When this straight line motion velocity gain is reduced by the relative velocity gain of the whole aggregate mass of the vehicles' moving reference frame; then the quadratic progression of the motivating force is still outperforming /outrunning the reduction in transfer mass straight line displacement velocity gain, still propelling the vehicle forward at a reduced rate. The underlying principle/reason is again the closed loop energy feedback described by the basic feedback formula for formula #8:

Output, from, feedback = Input, into, feedback / (1 + Transfer. function)
Reference: Kurt Gieck, section T9

The "Transfer function" is a sum/ratio of the vehicles' aggregate inertial mass to the flywheel assembly straight line motion inertial mass. This means the IP energy drive output into the aggregate vehicle mass can NEVER be absolute ZERO, as long as the cyclic rotor initial Input energy potential is superior and as long as the superior rotor kinetic energy is acquired independently to the vehicles' aggregate inertial mass. This principle of distribution of the propulsion energy magnitude on the basis of feedback is presented in Fig.6 and Fig.7. The next chapter is using an analysis procedure congruent with feedback analysis in electro dynamic circuitry and control engineering.

Fig.6
Device Reference Frame Motion Proof

$$E_{Effective} = m * r^2 * (1/2(\,\omega_a - \omega_b)^2 * 2/\pi$$

Available effective rotational Kinetic Energy

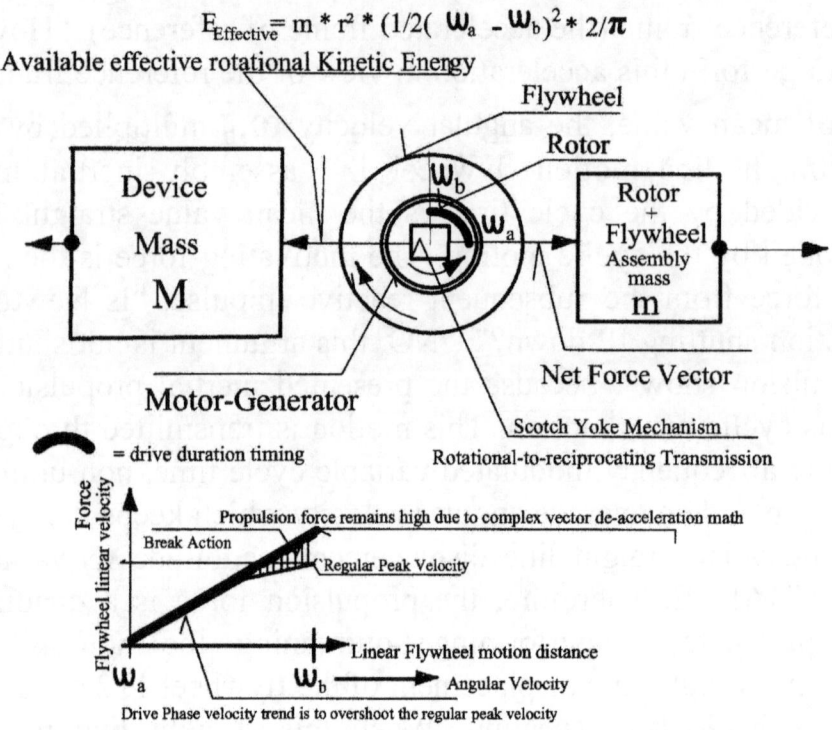

Flywheel

Rotor

Device Mass **M**

ω_b

ω_a

Rotor + Flywheel Assembly mass **m**

Net Force Vector

Motor-Generator

Scotch Yoke Mechanism

Rotational-to-reciprocating Transmission

= drive duration timing

Force / Flywheel linear velocity

Propulsion force remains high due to complex vector de-acceleration math

Break Action

Regular Peak Velocity

Linear Flywheel motion distance

ω_a ω_b → Angular Velocity

Drive Phase velocity trend is to overshoot the regular peak velocity

Reference Frame Motion Correction Logic

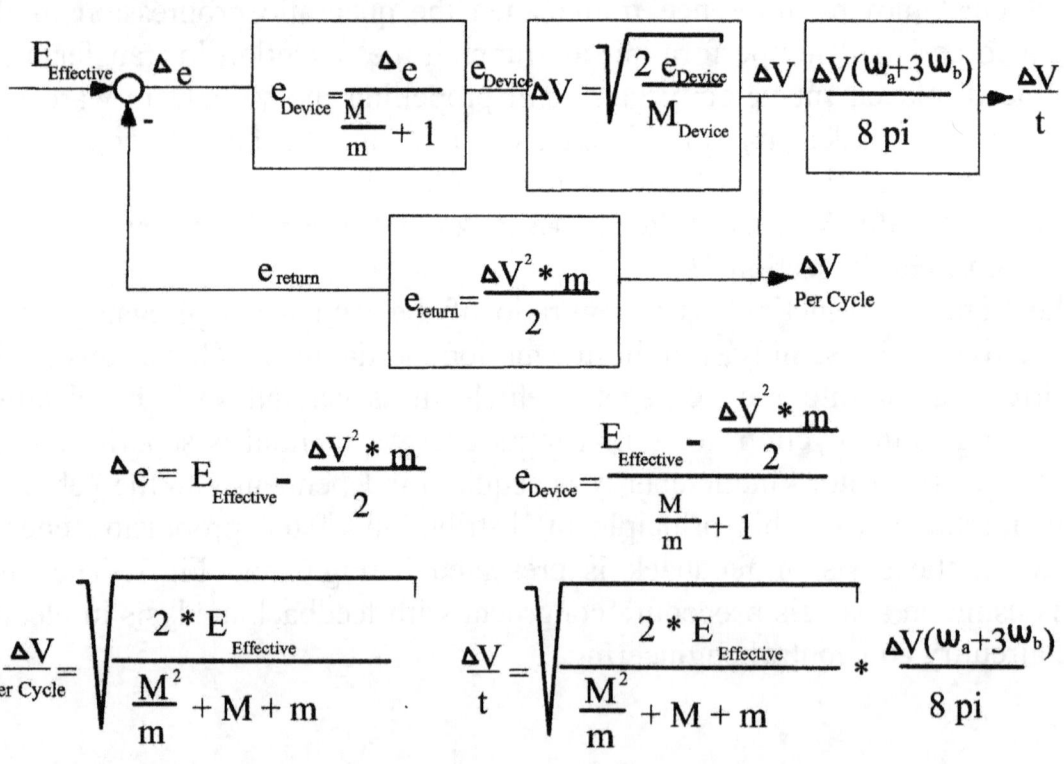

$E_{Effective}$ Δe

$$e_{Device} = \cfrac{\Delta e}{\cfrac{M}{m} + 1}$$

e_{Device} ΔV

$$\Delta V = \sqrt{\frac{2\,e_{Device}}{M_{Device}}}$$

ΔV

$$\frac{\Delta V(\omega_a + 3\omega_b)}{8\,pi}$$

$$\frac{\Delta V}{t}$$

e_{return}

$$e_{return} = \frac{\Delta V^2 * m}{2}$$

$$\Delta V \text{ Per Cycle}$$

$$\Delta e = E_{Effective} - \frac{\Delta V^2 * m}{2}$$

$$e_{Device} = \cfrac{E_{Effective} - \cfrac{\Delta V^2 * m}{2}}{\cfrac{M}{m} + 1}$$

$$\frac{\Delta V}{Per\ Cycle} = \sqrt{\frac{2 * E_{Effective}}{\cfrac{M^2}{m} + M + m}}$$

$$\frac{\Delta V}{t} = \sqrt{\frac{2 * E_{Effective}}{\cfrac{M^2}{m} + M + m}} * \frac{\Delta V(\omega_a + 3\omega_b)}{8\,pi}$$

Fig.7

Proof reference frame motion and gravitational pull

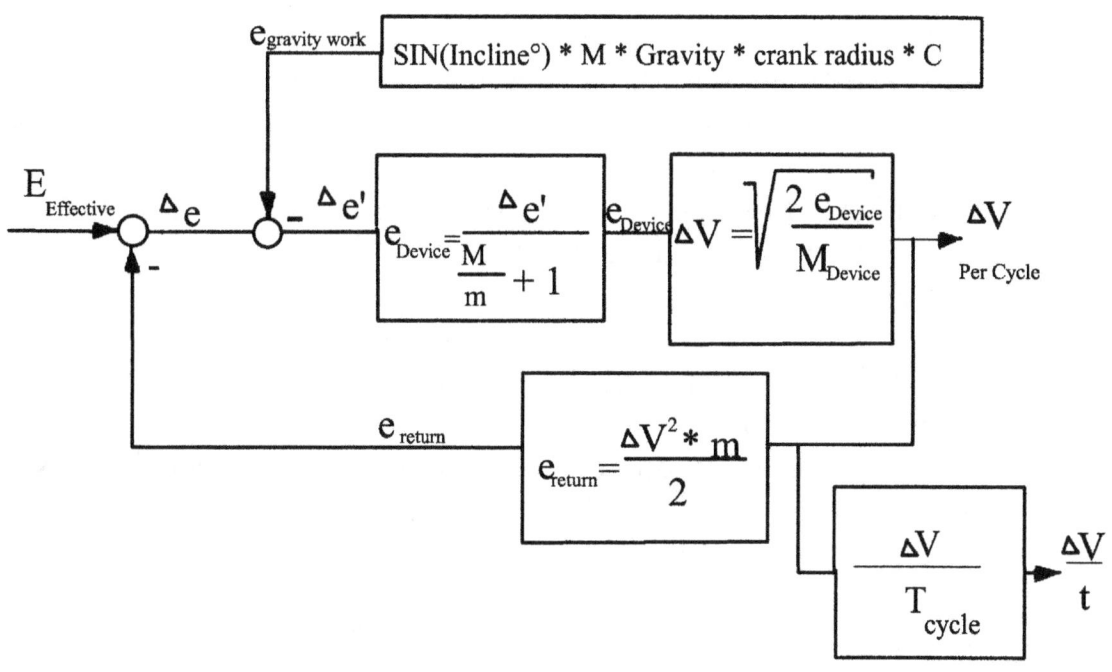

$e_{\text{gravity work}}$ = SIN(Incline°) * M * Gravity * crank radius * C

$E_{\text{Effective}}$ → Δe → $\Delta e'$

$e_{\text{Device}} = \dfrac{\Delta e'}{\dfrac{M}{m} + 1}$

e_{Device} → $\Delta V = \sqrt{\dfrac{2\, e_{\text{Device}}}{M_{\text{Device}}}}$ → ΔV Per Cycle

e_{return} $e_{\text{return}} = \dfrac{\Delta V^2 * m}{2}$

$\dfrac{\Delta V}{T_{\text{cycle}}}$ → $\dfrac{\Delta V}{t}$

$$C = \text{Drive phase to Idle phase time ratio} = \frac{t_1}{t_2 + t_3 + t_4}$$

$$\Delta e' = E_{\text{Effective}} - \frac{\Delta V^2 * m}{2} C$$

The Inertial Propulsion Formula

$$\Delta V_{\text{Per Cycle}} = \sqrt{\frac{2 * E_{\text{Effective}} - 4 * \text{SIN(Incline°)} * M * \text{Gravity} * \text{radius} * C}{\dfrac{M^2}{m} + M + m}}$$

$$\text{SIN(Incline°)} = \frac{2 * E_{\text{Effective}} - \Delta V^2 \left[\dfrac{M^2}{m} + M + m \right]}{4 * M * \text{Gravity} * \text{crank radius} * C}$$

Propulsion Energy efficiency and future potential of the presented IP technology

With all the presented facts it is certain that Inertial Propulsion does exist. However, what energy efficiency can be expected by analyzing the presented formulas?

Every 1 kgf,meter=0.0023Kcal of cyclic oscillating energy circulation, which comprises a 1 meter long ideal spring compressed by the weight of a 1 kg mass, is approximately only able to generate 0.02 kgf seconds or 0.20 Newton seconds of self-contained impulse. This represents approximately the impulse of throwing a 3 point skip with a nice sized pebble onto the surface of the local pond. The Inertial propulsion is, therefore, an energy-hog of **stored energy** circulation in the form of very high RPM speeds and extremely energetic short duration modulating pulses to accomplishing a viable IP drive.

The **energy circulation magnitude** should not be confused with energy drive magnitude which is the energy expended into the vehicle for every cycle which is driving the vehicular kinetic energy gain. No! A viable Inertial Propulsion Drive cannot be built with any old furnace motor. This illustrates the contrary technological challenges encountered to demonstrate the required pendulum test, a lifting operation within the gravitational field of almost g=10 meter/s; the only test the science community accepts as valid.

"Then in contrast, what exquisite technology have we accumulated to lift our space station into Earth Orbit or the Apollo Moon shot?". How does the Apollo moon rocket technology compare to the presented IP devices in terms of energy efficiency? Because very large volumes of energetic super-heated gases are expelled by rocket engines and lost in space.

The **THRUST MAGNITUDE** per same energy magnitude consumed by the presented IP drive is

proven to be <u>12</u> times <u>more</u> than the Saturn rocket.

The Apollo Saturn first stage moon rocket had energy consumption 4.3 Kwatt per Newton force thrust while the presented Inertial Propulsion device technology has

ONLY an energy consumption of 0.350 Kwatt per Newton force

Thrust exerted. In contrast, direct wheel drive car type drive efficiency is probably not possible with IP technology.

The presented IP drive consumes electricity which can be generated by Radioisotope Generators having 80 years of electricity supply; this allows a larger energy supply magnitude over time of the voyage than the very short energy burst of the very large chemical rocket employed today.

When considering a thermos-nuclear fuel, then the total fuel energy capacity is the nuclear fuel + the vehicular kinetic energy within that fuel,

$$E = mc^2 + \tfrac{1}{2}m_{,fuel}\, V^2_{,vehicle}.$$

This means: The kinetic energy accumulated within the fuel is returned to the IP drive. This theorem was introduced by the US Saturn Program rocket scientist Hermann Oberth in the 1950's. This is also called the **Oberth Principle** of space propulsion.

Now, we will consider the presented IP device attempting to accelerate past the velocity of light in view of the velocity limits set out with formula #1.5. The energy applied onto the vehicular inertial mass is:

$$E = m_{fission}\,(c^2 + \tfrac{1}{2}(c+V_{gain})^2);$$

wherein the term

$$\tfrac{1}{2}(c+V_{gain})^2$$

is the kinetic energy content of the delta mass $m_{fission}$ set free by the fission process.

The kinetic energy requirement for the acceleration past c is:

$$Ke_{new} = (M_{vehicle} + m_{orgin} - m_{fission})(c+V_{gain})^2$$

Solving the two terms for V_{gain} we get:

$$V_{gain} = c - c(m_{fission}(1+\tfrac{1}{2})/M_{vehicle} + m_{orgin} - m_{fission})^{\tfrac{1}{2}}$$

Accordingly, at the speed of light the IP vehicle has a velocity gain magnitude only dependent on the very large thermonuclear energy within the fission mass and the reference frame shift loss, because the kinetic energy is also additionally set free as an added energy supply.

The Radioisotope Thermonuclear Thermocouple array Power Generator , the RTG, has a 150 Watt power capacity, 2 kg mass, and has a 80 year power supply lifespan; this power-supply is suited to power the presented IP technology for a 80 Year time duration, driving it past the speed of light with help of planetary gravitational pull. Accordingly, Space propulsion has already vitality with the IP technology presented herein.

The Technology needed for a perfect Inertial Propulsion device:
1) A robust High yield electrical power supply.
2) High power to weight DC Drive motors in the multiple HP per 1 Kg power to weight ratio and with **zero** internal drag or cog losses.
3) High DC motors spin velocities in the multiple thousand RPM range and

High mechanical power response ratio to very fast very high amperage pulsed DC drive operation.

4) High stencil strength material mechanical parts constructed from carbon fiber material.

5) Friction less bearings.

The Final Analysis of the Reality of Operation applying to the combined effort inertial propulsion Device

The DYNAMIC Force Impulse of the Drive Phase is larger than the Force Impulse of the Accumulation and the Idle Phase; because the Drive Phase has a larger initial and also a larger average Impact Rotor (flywheel #1) angular velocity, which is a larger kinetic energy content in a squared functional progression. The larger drive phase initial angular velocity potential is accomplished with mutual reciprocal and singular exertions between the impact rotor (flywheel 1) and the flywheel 2 dynamic **backrest type** inertial reluctance having independence of any other exertions.

The flywheel assembly force reflections directed into a straight line motion direction is in a quadratic progressing function of the angular rotor velocity. Therefore it generates a dynamic reaction-less propulsion Force Impulse on the Device. The internal generated thrust of the device is a function of the impact rotor angular velocity (RPM) magnitude. The larger the RPM of the impact rotor becomes; then, proportionally larger becomes the operational thrust, in a diminishing returns progression. The kinetic energy delivered by each individual cycle is limited by the ratio of the devices' operating component masses and component stencil strength.

Accordingly, the analysis of the Inertial Propulsion device must proceed first from the cyclic energy flow, then to the self-contained impulse derived from the energy flow magnitude. The notion that such a device violates any conservation principles has been proven **to be unfounded**.

Detailed description of engineering the MARK II Inertial Propulsion Device

Referring to Fig.8 and Fig.9 placed near the end of the book. The two figures present a mechanical representation of the self-contained propulsion device comprising pairs of flywheels, 1 and 2. Each pair is having parallel axial orientation, opposite direction of rotation and opposite alternating straight line reciprocal motion. The reciprocal straight line displacement motion of each flywheel has equal stroke length, equal peak repetitive straight line displacement velocities and differential magnitude of straight line displacement motion inertial rotor driven accelerations. The differential magnitude of the rotor driven accelerations within one complete cycle of the reciprocal motion represents the source of propulsion energy. The flywheels 1 and 2 have a rotational and straight line kinetic energy storage capacity for providing a dynamic inertial reluctance backrest for the propulsion of the device.

The opposite alternating straight line movement of the pair of flywheels accomplishes the averaging of propulsion forces and the cancelling of rotational moments. More pairs of flywheels employed within the device results in better averaging of the propulsion forces and fewer vibrations can be expected. The device can also operate with the pairs of flywheels moving in simultaneous alternating straight line motion, which propels the device more in individual strokes than in continuous motion. The opposite direction of rotation accomplishes the cancellation of rotational forces, which prevents the turning of the device around its axis. The turning action is used to steer the device by varying the rotational parameters of the flywheel drives. The pair of flywheels 1 and 2, each contain an integral motor-generator for providing kinetic energy magnitudes into the impact rotor 3 and 4 (the rotors are congruent with flywheel 1, 2 in the proof section).The impact rotors are contained within motor-generator housing 3A and 4A and have the purpose of absorbing and delivering rotational kinetic energy. The motor housing 3A, 3B are firmly embedded into each flywheel forming integral assemblies.

The motor-generator impact rotor has a rotational kinetic energy storage capacity for delivering a rotational impact momentum. These motor-generators can be of different types of technologies, for example, a pneumatic vane motor-pump or a hydraulic gear motor-pump. For illustration, an electrical motor-generator armature with the current carrying conductors and the flywheel mounted field magnets are shown. The side-wall of the flywheel 1 is cut open to reveal the motor-generator within the flywheel. The motor-generator supplies alternating kinetic energy pulses to the flywheel assemblies, causing the rotation and progressively changing alternating straight line movement. The progressively changing straight line movement of the flywheels is the source of dynamic back-rest for the unimpeded exertion of the kinetic propulsion energy, which is fully explained in Fig.10 to Fig.12. The supporting frame 5, of the propulsion device is cut away from all attachment points for unimpeded view of the active working elements.

The propulsion device further comprises two straight line guides 6 and 7, which give each flywheel assembly a substantial straight line freedom of movement in the direction of vehicular travel. For the present design, swing-arms 6 and 7 are depicted, but many other technologies are suitable to guide the flywheels in a straight line motion. The swing-arms contain flywheels 1 and 2 on the moveable wrist-end, and pivot at the socket-end 8 and 9. The flywheels 1 and 2 rotate around the central shaft 10 and 11, by means of a set of rotational bearings 12 and 13, while the integral motor-generator impact rotor is firmly mounted co-centrically onto the central shafts 10 and 11.

The central shaft is contained on the wrist-end of the swing-arm by means of an additional set of rotational bearings 14 and 15, allowing the flywheels and the motor-generator rotor complete free-wheeling freedom of rotation in relation to the supporting frame of the device. The propulsion device further comprises pairs of rotational-to-reciprocating transmissions. The transmission type used by the present invention is the complimentary cam 16 and 17 and two cam followers 18 and 19. Many different mechanical constructs can be adopted as rotational-to-reciprocating transmissions, for example: A crank and connection rod, a scotch yoke mechanism or a hydraulic pump and cylinder, to mention a few.

The complementary cam 16 and 17 are firmly mounted onto each central shaft and have an ex-centrically contour in relation to the flywheel assemblies. The two cam followers 18 and 19 are mounted to the device frame. The rotational cams and the cam-followers motivate alternating opposing straight line movement of the flywheels.

The cam and cam followers convert the rotation and torque of the motor-generator impact rotor into progressively changing reciprocal straight line motion and straight line forces and provide a progressively increasing straight line, de-accelerated motion, in both directions of the reciprocal motion. The reciprocal motion delivered by the cam and cam followers has a preferred stroke length in fractions of the diameter of the motor-generator rotor diameter and as short as permissible cycle time. This is for maximizing the drive performance. The cam and cam-followers further provide a differential and reciprocal kinetic energy feed path, from the motor-generator rotor to the straight line motion inertial reluctances' of the flywheel assemblies. This feed path represents a negative feedback loop, reciprocally feeding and reducing the rotational kinetic energy of the motor-generator rotor into the straight line kinetic energy of the flywheel assembly. The feedback loop is feeding and depleting to zero the straight line kinetic energy of the flywheel into the rotational kinetic energy of the motor-generator rotor for the purpose of delivering a straight line to rotation coupled motion.

The straight line motion kinetic energy output of the rotational to reciprocating transmission is represented by the cam-followers 18 and 19. The cam-followers are mounted on the supporting frame 5, perpendicular to the flywheels axis. They act as the kinetic energy output path, acting against the device, which represents the entrance point of propulsion energy into the device. The straight line to rotation coupled motion of the pair of flywheels is delivering acceleration and de-acceleration forces against the cam- followers. The accelerations are caused by the cyclic flow of the kinetic energy, which flows between rotational and straight line kinetic energy. The combined sum of the propulsion forces are zero, if the combined motion of the pair of flywheels have no new kinetic energy introduced by the system. When new additional kinetic energy is induced, then additional acceleration forces are generated, which is the source of propulsion energy.

The propulsion device comprises a power-supply and a logic control 20, which contains the logic control that times and maximizes the efficiency of the working components. For the simplest form of the device, power cam switch operator 21 is mounted respectively, onto central shafts 10 and 11, and is able to supply timed power drive pulses to the motor-generator. Such an arrangement is set for one single set of operational parameters, for fast motion or stall hovering, but not for both at the same time. For complete automatic control, a computer logic algorithm must be used. The logic control has a command and control input 20 and is used for speed and directional control of the device. The method of directional control is accomplished with the differential variation of the duration and angle parameters of the motor-generator drive pulses. Power cam switches 23 and 24, supply power pulses to the motor-generator. For the simplest form of the device, break-shoes 25 and 26 are used for absorbing excess rotational flywheel kinetic energy from the flywheels 1 and 2. The break-shoes operate with a short momentary touch break action at approximately 265°-285°. The two break-shoe actions are rotational and reciprocal actions; therefore, do not impede the propulsion of the device. The break-shoes are mounted on swing-arms 27 and 28 which in turn are mounted on pivot points 29 and 30. The break-shoes are forced against the flywheel 1 rim with spring 31. The break-shoes can be of a variety of technology For example: Automotive type break-shoe material, electrical eddy-current technology, electro-magnetic poles technology or pneumatic type technology. The break-shoe actuator 32 is used for varying the position of the break-shoes.The variance of the position causes more or less demand for kinetic energy reduction in the flywheels 1 and 2, by varying effectively the duration of the break-shoe contact action. Any unwanted oscillating rotation motion and gyration of the propulsion device, which is induced by the break shoe action, is arrested with variable eccentric flywheel 33 and the rotational drive 34. The rotational drive can be of a variety of technologies, including a synchronous motor. The cam switches must be of proxy-sensor-drivers technology for angular velocity over 100 RPM. The device is switched on with switch 35. Idle power and drive power is selected with switch 36. The rheostat 37 is for adjusting the drive power balance. The electrical choke 38 is for shaping the alternating drive pulses. The arrow 39 indicates the direction of vehicular travel. For illustrating the dimension of present typical combined effort inertial drive, dimensions are given for the flywheel D=200mm, for the rotor d=60 mm, 0.7 kg mass for the flywheel and 0.3 kg for the rotor. The mass moment of inertia of the flywheel is $0.0022Kgm^2$. The mass moment of the rotor is $0.0005kgm^2$. The stroke of the cam is 0.01meter for every ¼ turn. The ratio of one flywheel assembly in comparison to the second flywheel assembly, plus the frame, is one to three and the total device mass is 4Kg.

Fig. 10 depicts the graph of the motor-generator alternating drive pulses in correlation to the angular rotation of the motor-generator 3 in Fig. 1. The graph depicts the alternating drive pulses for the motor-generator rotor. The drive pulses have a positive and negative swing. The positive drive pulse swing is the source of kinetic propulsion energy and the negative drive pulse swing absorbs the unused kinetic energy. The difference between the positive drive pulse swing and the negative drive pulse swing represents the amount of kinetic energy induced into the propulsion of the device. The motor-generator impact rotor positive drive pulses start between 20°-90°, which drives and accelerates the flywheel 1 in the clockwise direction and drives the motor-generator impact rotor 3, in the counter-clockwise direction. The reciprocal rotational acceleration drive of the motor-generator impact rotor against the rotational inertial reluctance of the flywheel represents the principle of equal reaction to the action of the flywheels' free-wheeling inertial acceleration. During the positive drive angular acceleration, rotational kinetic energy is accumulated into the motor-generator impact rotors 3 and 4, which is subsequently used for the propulsion of the device. The positive drive pulse duration is called the accumulation phase. The driving and straight line acceleration of the propulsion device is accomplished by the relative de-acceleration of the angular velocity of flywheels 1 and 2 and the accompanying relative de-acceleration of the angular velocity of the motor-generator impact rotors 3 and 4. The driving and acceleration of the device occurs between 90°-270°, which accelerates the straight line inertia of the flywheels assemblies in opposite direction of vehicular travel at an increased rate, which induces kinetic energy into the device and drives and accelerates the device forward. The device drive phase is an impact by separation, effectively converting the accumulated rotational kinetic energy of the motor-generator impact rotor into straight line kinetic energy of the device applying to formula #8. The rate of de-acceleration of the motor-generator rotor, during the drive phase, is a measurement for the amount of kinetic energy invested into the device and is used as an input parameter by the logic control. The drive phase employs a negative motor-generator rotor drive pulse to remove the unused rotational kinetic energy, which was not induced into the device, and also restores the unused kinetic energy back into the power-supply. The logic control meters the negative drive pulse to ensure that the peak straight line velocity of the reciprocal motion of the flywheels remain constant and the flywheel assembly straight line acceleration remains at a minimum level for maintaining a continuing positive drive force up to the point of maximum straight line velocity. The logic control also meters the negative drive pulse which determines the optimum intensity and start time; because kinetic energy must remain in the motor-generator impact rotor 3, to complete the rotational cycle at the average velocity. Therefore, the differential magnitude of kinetic energy between the positive drive pulse and the negative drive pulse is the total kinetic energy invested into the motion of the vehicle. The device complies with the principle of conservation of momentum. In case the device is held at rest, then no kinetic energy is transferred into the inertia of the device. In case of the device motion stall the angular velocity of the motor-generator rotor will not decrease from 90° to 270° and the maximum propulsion force is exerted against the kinetic energy output cam follower. This is because, during a stall, the maximum straight line acceleration of the flywheel assembly occurs and most (except friction losses) of the kinetic energy induced by the positive drive pulse is then removed by the negative drive pulse. In case the negative drive pulse is not present and the positive drive pulse remains at full capacity, then the kinetic energy and hence the angular velocity of the motor-generator rotor will increase accordingly. The device is complying with the principle of conservation of kinetic energy. Fig. 10 and 12 depicts the timing method used and employs a cam with opposing cam followers. Fig. 11 plots the straight line velocity of the flywheel assemblies 1 and 2 in comparison to the rotation of the complimentary cam in degrees. It depicts the flywheel assembly straight line distance traversed, the resulting cycle time and the differential half cycle velocities. This differential of straight line velocity is the source of propulsion power for the device. Because of the before mentioned velocity-differential of the devices mechanical motions from 0-180° to 180°-360°, a method of multiple integral flywheel motor-generator assemblies must work in a timed union to accomplish a continuous and seamless propulsion drive. Fig.12 plots the motor generator rotor and flywheel angular velocities. The cycle frequency of the impact rotor at approximately 960RPM rotational speed is paltry. This cycle frequency can be tuned up to 9000RPM using new technology components and delivers a 10 fold increase in total power into the propulsion of the device. It becomes compellingly apparent, from the analysis of the diagrams in Fig.10 to Fig.12, that the method of the inertial propulsion of the device is the progressive, reciprocal and alternate building and depletion of the kinetic energy levels in the straight line and rotational flywheel assemblies' masses. The kinetic energy flow is superimposed with induction and deduction of fresh kinetic energy with the resultant progressive kinetic energy flow, which in turn generates the progressive net impulse magnitudes for the propulsion of the device. Page 104 and105 depicts the calculation footprint of the self- contained forces of the device when the device is in motion and held at rest. The calculation method for the stall condition is different than the energy distribution method used when the device is in motion; because no kinetic energy flows into the devices' straight line motion kinetic energy, then the available rotor kinetic energy is only distributed into the flywheel assemblies' mass straight line vector component. The stall force is also the force the device would force against the gravity of the earth and hover without changing position. To obtain the earth gravity stall force a 10HP motor would required within each flywheel, delivering 7400 Watt of power each and have a unit mass of 0.300Kg and an average RPM of 20000.

Today we are not within reach of a 1 Hp motor at a weight of 0.3 Kg of the new Core-less, Iron-less DC motor technologies needed for the presented Inertial propulsion Drive, but this technology might be available in the near future.

The Mark III device, CIP US Patent Application#: 14/120031
The Mark II device has a disadvantage in energy efficiency; this is because of the touch break flywheel #2 breaking action. Furthermore, the Mark II device employs an expensive logic controller making it a rather expensive hardware and software device and the cam and cam follower design is furthermore expensive in view of friction losses and hardware cost. All these design disadvantages are fixed with the mark III design employing:

1) Solid state sensors.
2) Solid state motor generator drivers.
3) Manual adjustable Controller.
4) Friction-less Push rod followers replace the cam-followers.
5) Circulating permanent magnet engaging with stationary electro-magnetic coils recycle the excess flywheel angular kinetic energy into the power supply.

The math and physics formulas remain the same for the Marl III design.

Fig.8

Side View of the Device

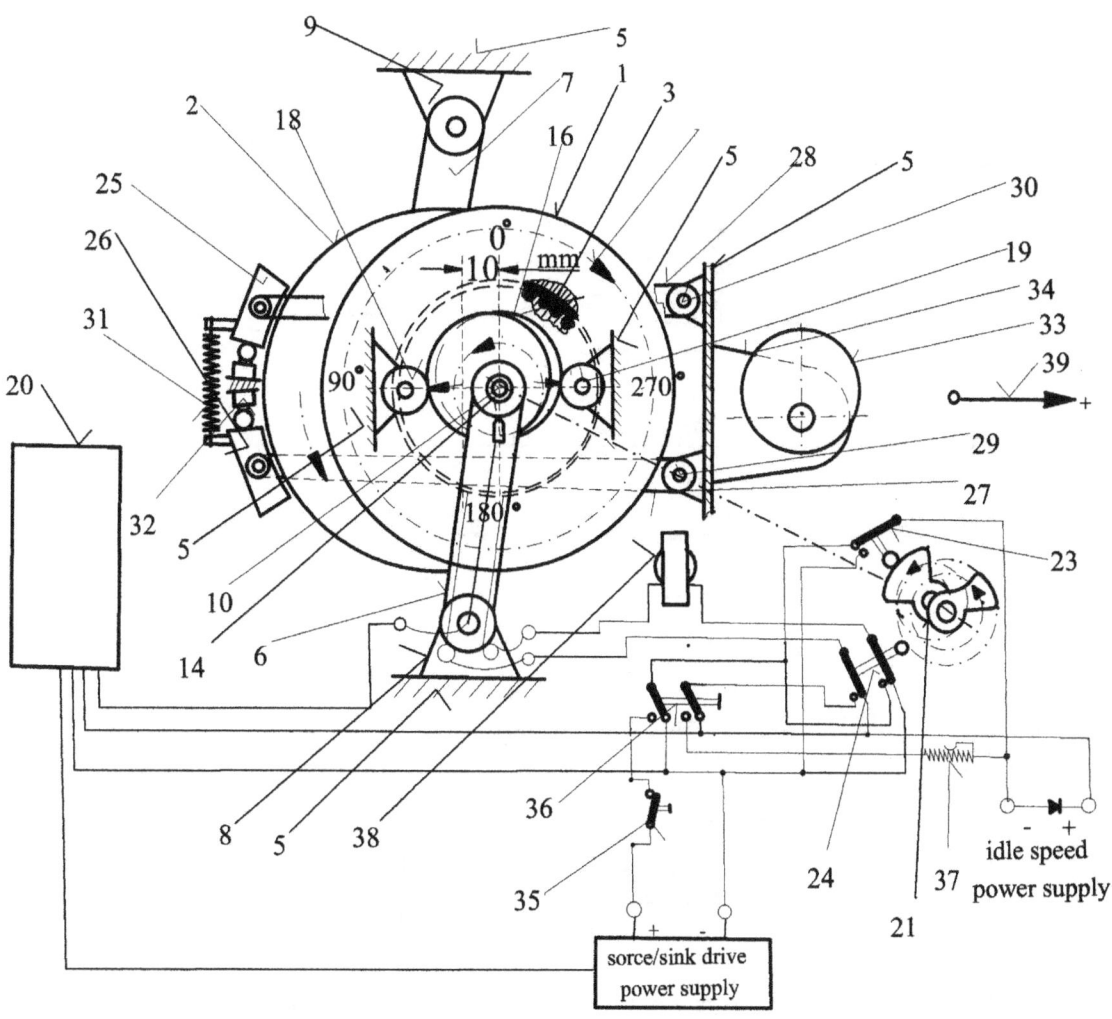

Fig.9

Top View of the Device

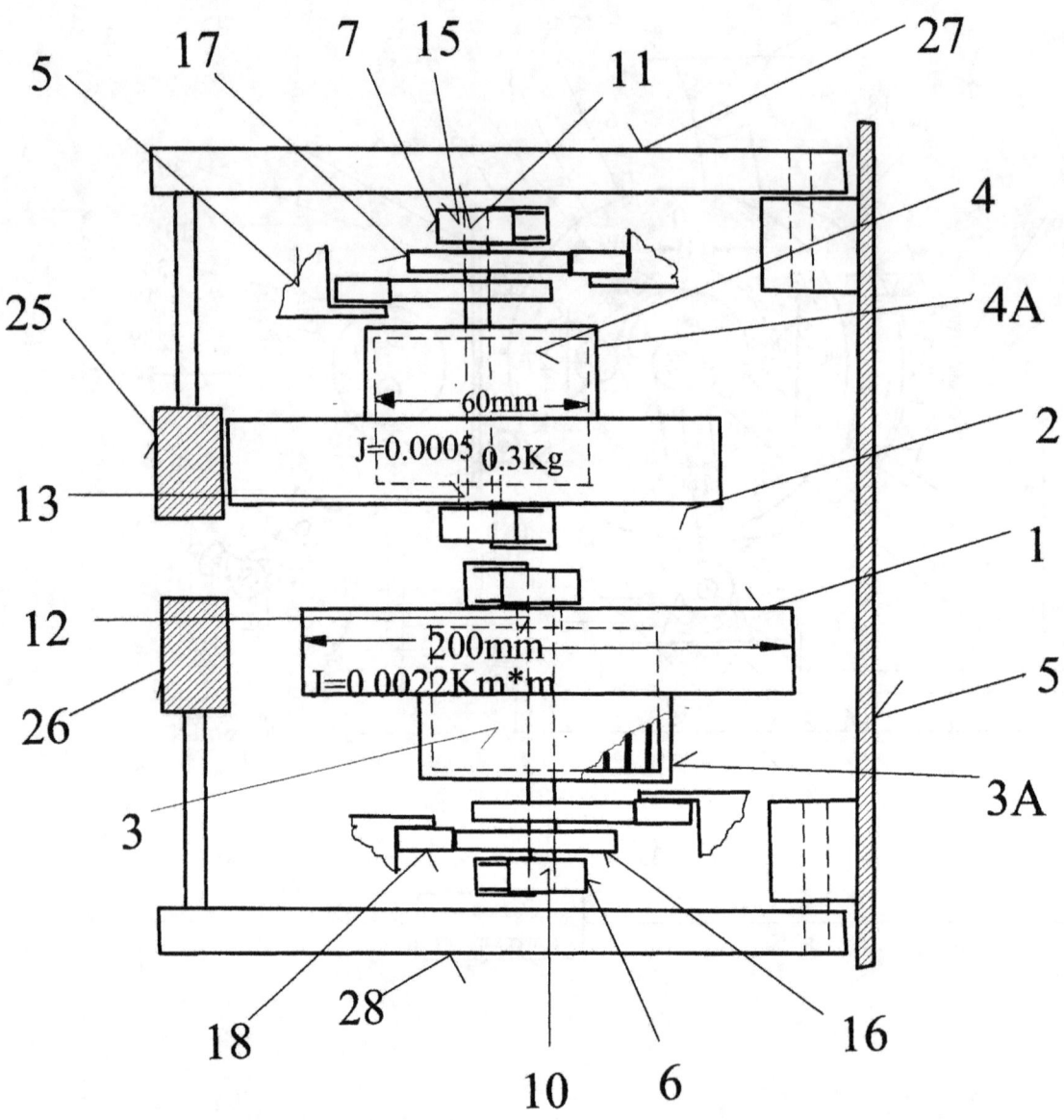

From the motor-generator drive pulses footprint Fig.10 we can extrapolate the straight line velocity in Fig.11 and Fig.12 in the following page.

Fig.10

Energy footprint of the motor-generator

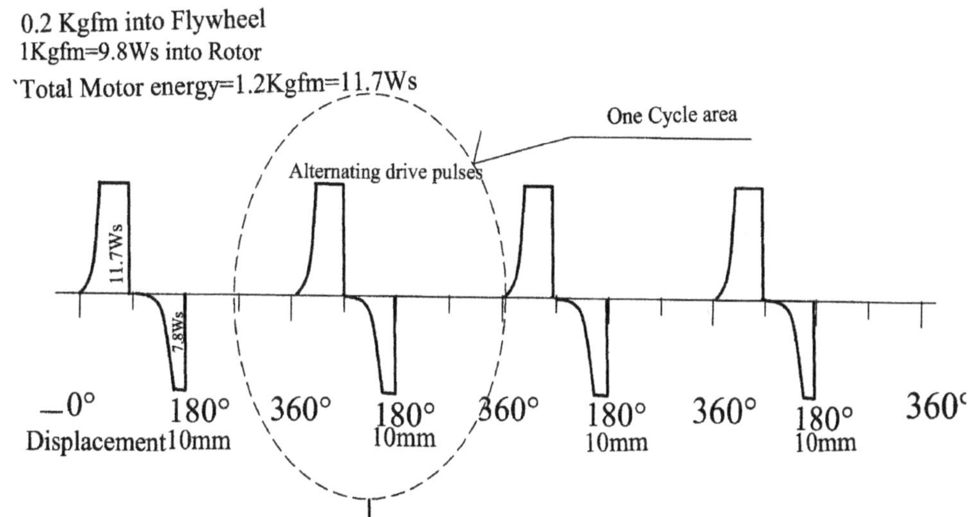

0.2 Kgfm into Flywheel
1Kgfm=9.8Ws into Rotor
`Total Motor energy=1.2Kgfm=11.7Ws

One cycle region of intrest

Fig.11

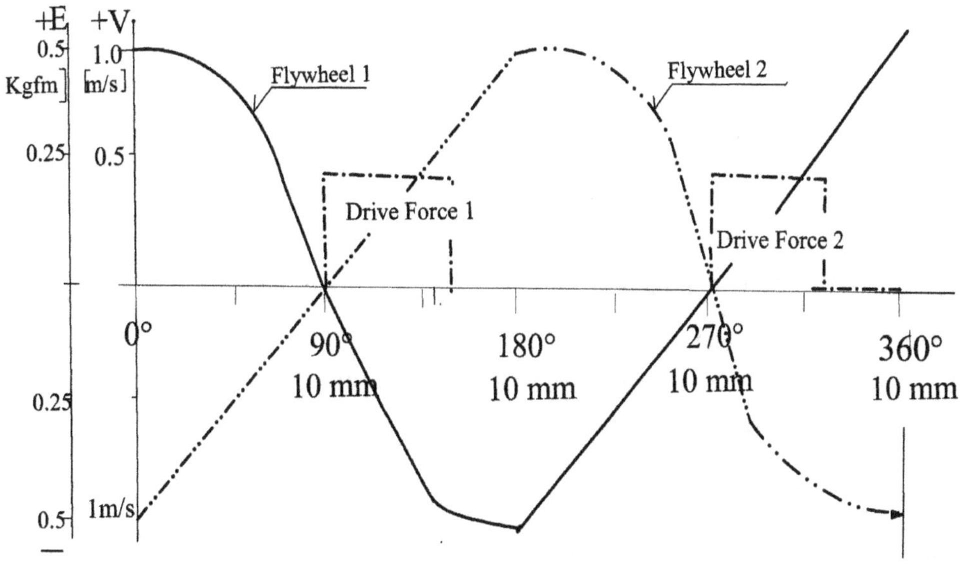

Fig. 12

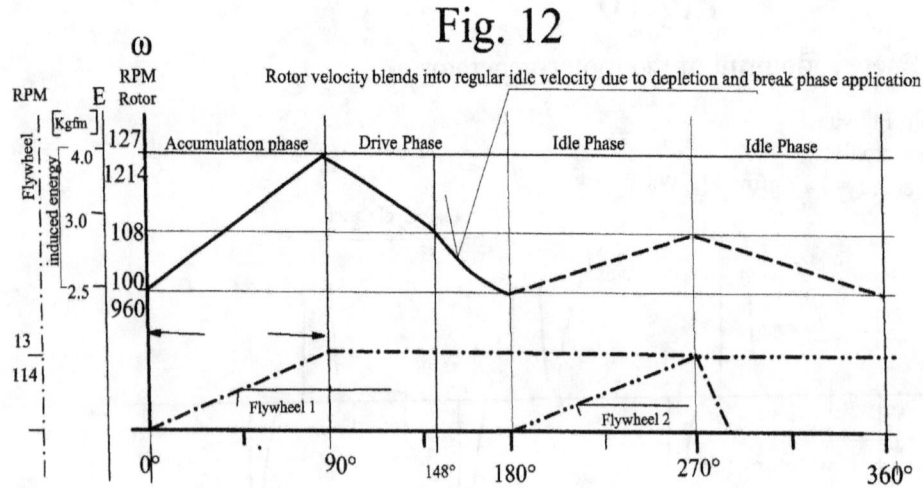

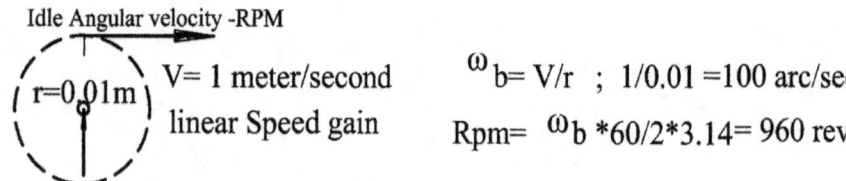

Calculation Footprint

Rotor Mass Moment of Inertia $J = m/2 * r * r$; 0.3 Kg / 2 * 0.06 * 0.06 meter = 0.0005Kgm*m

Flyweel Mass Moment of Inertia: $J = m/2 * (R*R + r*r)$; 0.7 Kg /2 (0.1 * 0.1 - 0.06*0.06) = 0.0022kgm

Idle Angular velocity -RPM

V= 1 meter/second
linear Speed gain

$\omega_b = V/r$; 1/0.01 = 100 arc/sec

Rpm = ω_b *60/2*3.14 = 960 revolution/min

Idle cycle Duration

T = distance/Velocity ; $2*3.14/\omega$ average = 0.064 seconds

Linear Kinetic Energy of the Flywheel assembly

Flywheel assembly mass = 0.3+0.7 = 1Kg

Idle kintic energy at 0°: E = 1Kg/2*1*1meter/secont = 0.5Kgfm

Rotor kinetic energy at 0° e = 0.0005/2*100*100 = 2.5Kgfm

Kinetic energy induced during the accumulation Phase: = 1Kgfm

Rotor Angular velocity at 90°: $\omega_a = \sqrt{2*4/0.0005}$ = 127 rad/second

RPM = 1214 revolution/min

Calculating the constants and phase time durations

2* flywheel assembly mass=2kg

Device frame mass=2kg

total device mass=4kg

impact rotor mass moment =0.0005kgs²m

radius of gyration=0.01meter

natural decay multiplier for the drive-phase=

$C_1=1/(0.001*0.001/0.0005*4+0.001*0.001/0.0005+1)$

$C_1=0.8771$

natural energy conservation multiplier for the idle phases=

$C_2=(0.01*0.01/(0.01*0.01+0.0005)+1)$

$C_2=1.08$

$C_3=-2.00$

$t_1=\pi/(127+100)=0.01384$ seconds

$t_2=\pi/(100+100*1.08)=0.0151$ seconds

$t_3=\pi/(100*1.08*0.8771+127)=0.0155$ seconds

$t_4=\pi/(100*1.08*0.8771+127)=0.0142$ seconds

T, total=0.0587 seconds total cycle time

initial analysis using formula #11

1N=0.102Kg,force

#11) Impulse = mass, transferred * radius * (1/2(ω,a-ω,c))

p=0.102*1*0.01*0.5(127-108)=0.009 Kgf,second per flywheel

P=0.088 Newton,second per flywheel per cycle

p=0.018 Kgf, seconds per two flywheels per cycle

P=0.17 Newton,second per two flywheels per cycle

Mathematically Exact Impulse=

$$P = \text{flywheel mass} * \text{radius} * 1/2(\omega,a+\omega,b(1+C_3))$$

$$P = 0.102*1*0.01*0.5(127+100(1+(-2.0)))=0.0138 \text{Kgf, second per}$$

flywheel: This proofs the minimum validity of formula
#11returning a minimum impulse value.

Internal force is diluted by the total cycle time =
$0.0138\text{Kgfseconds}/0.0587\text{seconds}=0.235\text{Kgf}$ per flywheel

Internal Force$=2.304$ Newton per flywheel

is the net motivating internal self-contained Force

Work-Energy, effective, net$=E_{net}=$Force, self-contained,
*Force, internal, self-contained *radius$=$
$0.235*0.01=0.00235\text{Kgfm}$, 0.023 joules Energy per
flywheel.

Energy for two flywheels$=0.0047\text{Kgfm}$,

Energy for two flywheels in joules, $E=0.0461$ joules

Velocity, gain per cycle for two flywheels:

$$V_d = ((2*E/((4^2/1)+4+1)))=0.066 \text{ meter/second, per cycle}$$

for a two flywheel device.

Acceleration of the device on a level surface:

0.066meter per sec/0.0587second$=1.124$ meter/second^2

1.14/10 of Gravity

Maximum climbing ability:

$$C=t_1/t_2+t_3+t_4=0.309$$

$\text{Sin} =(2*2*0.023-0)/(4*4*9.8*0.01*0.309)=10°$ stall angle.

The Device is able to climb a 9° ramp at very slow speed.

**This is the Author's Mark I Inertial Propulsion
Devices'quantifiable-physical experimentalproofproducing a steady
30 gram self-contained thrust on a pendulum from 100Watt energy
flow. The Mark I first ran in October 2005.**

Within this picture of the experiment it is observable that the axis of the impact rotor-flywheel spin is parallel to the axis of the pendulum. The experiment where the axis of the spin is perpendicular to the pendulum axis is producing **zero** lift of the pendulum. This phenomena can be attributed to the stiff gyro action of the fast spinning Top, in respect to the head over heal rotation of its axis. This means that such a device would have a hard time to do a looping maneuver in respect to a vertical axis. Thisphenomenonwas particularly noticeable in the Gyrobus where the flywheel axis was also mounted vertically for passenger space convenience. The Gyrobus tried to remain in the horizontal position on Swiss mountain slopes and caused a strange kneeing effect of the bus suspension.

The pendulum test device is an 8 foot (2.5m) long compound pendulum which is tuned to display a pointer reflection of 1 mm per 1 gram-force of thrust. Next is a better picture of the pendulum test.

An enhanced picture of the Mark 1 model pendulum test

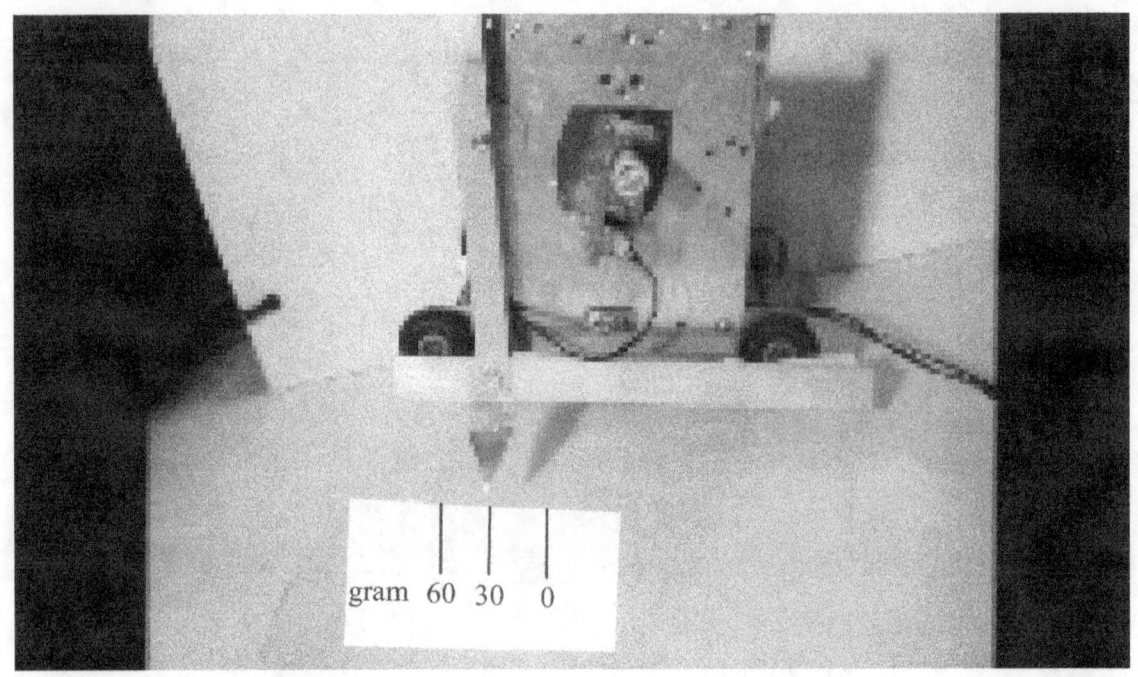

gram 60 30 0

This is the motor-generator armature voltage footprint of the Author's first successful IP device Mark 1 model.

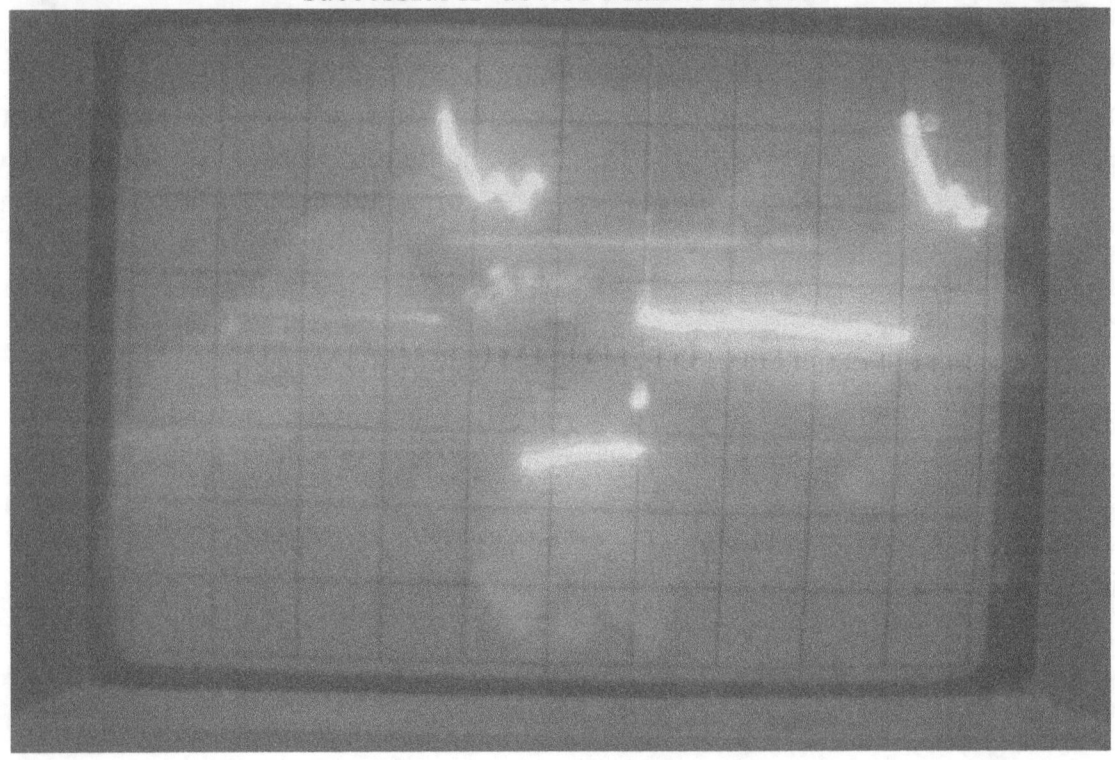

The scope time base is 20ms per division and the vertical is 20 volts per division.

The presented scope trace represents the superimposed motor-generator EMF voltage combined with the alternating drive voltage. The generator EMF voltage is proportional to the RPM magnitude. EMF stands for electric motor force.

The accumulation Phase is the first positive pulse from the left. Itis 24ms long and has a peak voltage of 62 volts applied to a 7Ω motor-generator armature and has a generator EMF voltage of 8 volts at the beginning of the accumulation Phase.

The Drive Phase is the negative pulse and is 27ms long. Ithas a peak voltage of 30 volts which is applied to the 7Ω motor-generator armature and has anopposing generator EMF of 12 volts.

The two Idle Phases are atotal time of 65ms. They have a small decline in angular speed due to mechanical and magnetic friction. This is indicated by the decline of the scope trace by approximately 4 Volts.
The total cycle time is 116ms.

The motor-generator average RPM is=517 revolution/minute. This must be viewed as the minimumaverage RPM to accomplish IP.
Furthermore, the RPM speed of 517 per minute is the balance point where the induced Accumulation Phase kinetic energy is in balance with the mechanical-magnetic friction energy consumption. This magnitude of RPM cannotbe exceeded with the fixed 100 Watt accumulation phase power.

The RMS wattage reading of the 62volts accumulation Phase power supply is 99 continuous Watt and is a net reading taken during the device run.
The calculation to convert the ¼ cycle phase durations to angular speed is:

$$\omega_{¼}=2\pi/t*4$$

The average accumulation phase angular speed is=$2\pi*1000/(24*4)$=65 1/s.

The average Drive Phase angular speed is=$2\pi*1000/(27*4)$=58.2 1/s

The average Idle Phase angular speed is= $2\pi*1000/(64*2)$=50 1/s

The total decline of the idle angular speed due to friction is 1/5 of the average idle angular speed. Therefore, the average angular speed is also the ω,c =50 1/s.At the beginning of the idle phase the angular speed is =ω,b=52 1/s, at the end of the idle phase the angular speed is 47 1/s.

The slight rise of angular speed from ω,b to ω,c indicates there is areasonable tuned mechanical oscillator resonant condition. Accordingly, we can assume ω,c is equal to ω,b angular speed=52 1/s.

175

The relationship of ω,a in relation to $\omega,b=52$ 1/s is:

$$\omega,a =(\pi*1000/27ms*) - \omega,b=65 \text{ 1/s}$$

This formula assumes an angular speed progression congruent with Fig.2. Accordingly, the peak angular speed is=ω,a=65 1/s

At the end of the Drive Phase,the angular speed is =ω,b=52 1/s. This is considering a drop of average Idle Phase angular speed of one 1/s, from 52 1/sdown to 47 1/s. This indicates that the idle phase consumes 11% of the supplied accumulation Phase power flow. The total waste power consumption for all phases, considering the mechanical and magnetic frictionpower consumption, is more than 24 continuous Watt energy flow. Further increasing the accumulation phase power pulse also increases the dissipative power in the form of heat. This represents a heat flow problem needing additional cooling.

Initial analysis using formula #11

Transfer mass =0.81kg. This includes the flywheel, motor-generator, impact rotor. Radius=8mm =0.008 m

#11) Impulse = mass, transferred * radius * (1/2(ω,a-ω,c))

0.81Kg * 0.008 meter (1/2(65-52))=**0.042 Newton-Second** Net impulses per cycle

This is a p=**4.3gram-force-second**Net Impulse per cycle

Dividing the Impulse per cycle by the total cycle time provides the continuous propulsion thrustgram-force:

Force, continuous=p4.3/t0.116 second=**37** gram-force-trust

This magnitude is a math based ideal quantity for perfect uniform angular speed changes. It assumes a steady increasing and decreasing angular speed. However, the very high initial Accumulation Phase current inrush of I=(U-EMF)/R=(62V-8V)/7Ω=**7.7** amperesrepresents an inrush of:416 Wattor more than 1/2 HP.Thiscausesthe power supply voltage to dip-slope down. This is observable in the scope picture and decreases the available accumulation power supply power flow. This dip-slip down is indicated by the 60 Hz charge pulses becoming visible in the accumulation phase drive pulse. The angular velocities depicted in Fig.2 in a straight linear line are depicted here like a sagging rope suspension bridge.

Furthermore, the friction of the pendulum pivot, the large internal mechanical/magnetic friction and the flywheel mechanical break shoefriction areconsidered as detrimental and we arriving at the real displayed / observed thrust force of **30 gram** of force.

This initial MarkI device first ran on a level surface in October 2005. The pendulum test with scope picture was performed inApril 2012. This proved the math and physics involved. Furthermore,it proved the large power flow requirement of an Inertial Propulsion device using conventional iron-core motor-generatorsandreveals the need to reduce the internal magnetic and mechanical friction.

The relationship between the Accumulation Phase and the Drive Phase power magnitudes is optimised to a perfect flow-timing condition with a manual operated rheostat. This indicates the need for a computer logic controlled optimization.

To illustrate the 30 Gram thrust achieved in relation to the motor-generator size used within the Mark I experiment, a picture of the motor is provided on the bottom of the page.

The presented Mark I test is not an attempt to achieve the world record of IP. It is to prove the viability and the math of IP.

Here is the physical motor size used by the Mark I device:

Here, we arrive at the completion of the analysis of Inertial Propulsion.
In view of the overwhelming proof in favor of the vitality of Inertial Propulsion may the Science Community be freed of any scepticism of Inertial Propulsion unless able to prove with a contrary formal scientific proof in validating the presented facts.

ABOUT THE AUTHOR

The author, Gottfried J. Gutsche has an education in Mechanical Control Engineering, Mechanical Design, Cybernetics, Electrical device design Engineering and Fortran-programming applying to the automatic control of electrical motors for automatic high speed inertial mass manipulation in factory automation. In depth, he studied automatic machine inertial mass motion control loop stability analysis and design in view of the 1960ties technology. Subsequently, he had the privilege to work 28 years in Computer data progressing technologies as a team member of an elite, highly skilled, stop the investment loss, fast enthralling pace, young dynamic men's game computer system problem diagnosticians; this positioned the author with an excellent view for witnessing the inside development of hyper complex computer system technologies from its mechanical-discrete transistor circuit technology infancy to very large scale electronic integrated circuitry reliability maturity; furthermore he witnessed the rapid technological development of operator manipulated magnetic tape information storage devices into autonomous robot fed central computer operated Mass Storage Systems using magnetic tape cartridge silos with up to 472 billion bytes of stored data maturity. Drawing from this rich technological development experience, he operated an automation machinery consulting business specializing in resolving stubborn ongoing factory automation problems, wherein the solution often boiled down to finding and resolving the many forms of stubborn energy flow restrictions. He has a combined total of more than 50 years of professional experience. The previous long work experience as a stop the investment loss project diagnostician honed the author to deliver fast, consistent very high quality analysis on difficult problems relating to inertial mass manipulation within machines. All illustrations and drawings are produced by the Author with a cad package. The author successfully self-prosecuted three completed patents referenced by 7 other patents as prior technology and has currently 5 patent applications.

At this point, the Author would like to thank all the people for their support and encouragement to produce this book, especially my wife Margaret.

Reference summary:

C. Huygens: "Motu Percussione: A. D. 1656",
 "Centrifuga: A. D. 1659",
"Oscillatorium: A. D. 1673"

Steve Hawking: "On Shoulders of Giants", ref. page 748

The Engineering reference: "Kurt Gieck Engineering Formulas 7Th Edition"; Ref. section L1-L9.

Mechanical reference for the continuing cycling Propulsion: "Kurt Gieck"; section L10, P10

 Schaum's: " 3000 Solved Physics Problems"; Ref. used for verifying mechanical examples.

Schaum's: "3000 solved problems in Physics"; Problem 4.15" for verifying the "Third Law in energy form"

For Calculus problems Schaum's 3000 Calculus problems: problem# 20.60, 31.26, 31.31.

DiStefano: "Complex control systems" Schaum's "Feedback and Control Systems"
.

Jean-Pierre Gazeau: "Group 24".

M. Browne: "Physics for science"

Heinrich Hertz: "Mechanics presented in a new form"

References from the internet:
www.UCSD/course/web.pages
www.wikipedia.org/Centipetal_force/
www.Farside.ph.utexas/Teaching/301/lecture/node89.html
www.electronics-tutorials.ws/accircuits/average-voltage.html
www.epi-eng.com

For Advance method of Dynamics by GW Housner at Caltech:
http://authors.library.caltech.edu/25023/

Internet Presentations:

Inertial Propulsion lecture presentation:
www.mindbites.com/series1278/lesson1
www.mindbites.com/series1278/lesson2/
www.mindbites.com/series1278/lesson3/
www.mindbites.com/series1278/lesson4/
www.mindbites.com/series1278/lesson5/
www.mindbites.com/series1278/lesson6/

Education Device pendulum test:
www.mindbites.com/series1278/lesson7/

Inertial Propulsion continuous cycling device:
www.mindbites.com/series1278/lesson8/
Video Presentation on youtube:
youtube, ggutsche1, P1300005
youtube, ggutsche1, P1100007
youtube, ggutsche1, P1090003
youtube, ggutsche1, PC130005
www.youtube.com/watch?v=3ZA1_uokiPs
https://www.youtube.com/watch?v=3ZA1_uokiPs

Notes: